SpringerBriefs in Applied Sciences and Technology

SpringerBriefs present concise summaries of cutting-edge research and practical applications across a wide spectrum of fields. Featuring compact volumes of 50–125 pages, the series covers a range of content from professional to academic.

Typical publications can be:

- A timely report of state-of-the art methods
- An introduction to or a manual for the application of mathematical or computer techniques
- A bridge between new research results, as published in journal articles
- A snapshot of a hot or emerging topic
- An in-depth case study
- A presentation of core concepts that students must understand in order to make independent contributions

SpringerBriefs are characterized by fast, global electronic dissemination, standard publishing contracts, standardized manuscript preparation and formatting guidelines, and expedited production schedules.

On the one hand, **SpringerBriefs in Applied Sciences and Technology** are devoted to the publication of fundamentals and applications within the different classical engineering disciplines as well as in interdisciplinary fields that recently emerged between these areas. On the other hand, as the boundary separating fundamental research and applied technology is more and more dissolving, this series is particularly open to trans-disciplinary topics between fundamental science and engineering.

Indexed by EI-Compendex, SCOPUS and Springerlink.

More information about this series at http://www.springer.com/series/8884

Luca Spiridigliozzi

Doped-Ceria Electrolytes

Synthesis, Sintering and Characterization

 Springer

Luca Spiridigliozzi
University of Cassino and Southern Lazio
Cassino, Frosinone, Italy

ISSN 2191-530X   ISSN 2191-5318   (electronic)
SpringerBriefs in Applied Sciences and Technology
ISBN 978-3-319-99394-2   ISBN 978-3-319-99395-9   (eBook)
https://doi.org/10.1007/978-3-319-99395-9

Library of Congress Control Number: 2018951697

This Springer imprint is published by the registered company Springer Nature Switzerland AG
The registered company address is: Gewerbestrasse 11, 6330 Cham, Switzerland

Contents

List of Figures

Chapter 1
Introduction

Over the last two decades, interest driven by the continuously increasing energy demand grew worldwide toward alternative power generation sources in order to lower the human carbon footprint and to improve the overall efficiency for electricity production.

Solid oxide fuel cells (SOFCs) are surely considered as one of the most valuable alternative sources to produce electricity with high efficiencies. In fact, fuel cell systems compared to the traditional ones can exhibit much higher yields, being not controlled by the Carnot yield of traditional thermal cycles. Moreover, such electrochemical devices can provide additional advantages in terms of reliability, fuel flexibility, low NO_x and SO_x emissions, noise emissions, etc. Up to about a decade ago, SOFCs have been designed to primarily operate at high-temperature ranges, i.e. 800–1000 °C, not only due to technological issues (i.e. components material requirements) but also to be used as internal hydrocarbon fuel reformers and valuable heat co-generators. In fact, high-temperature SOFCs can provide high-quality thermal energy and they can be easily coupled with gas turbines to significantly rise up the overall conversion efficiency of such combined power systems.

Anyway, high-temperature-related issues could not be underestimated in terms of materials, seals and components design, cells, and stack degradation. Because of these disadvantages, a new class of SOFCs operating at intermediate temperatures, i.e. 500–800 °C, has been considered during the last few years as the second-generation SOFCs. Such operating temperature decrease leads to simplified thermal management, easier startup, and cool-down steps and, more importantly, to a broader set of materials usage. Unfortunately, lower temperatures make electrolyte conductivity and electrodes kinetics drastically decrease. Those drawbacks gave rise to an extensive investigation of novel cell materials and design to be used in the intermediate-temperature SOFCs (IT-SOFCs).

© The Author(s), under exclusive license to Springer Nature Switzerland AG 2018
L. Spiridigliozzi, *Doped-Ceria Electrolytes*, SpringerBriefs in Applied Sciences and Technology, https://doi.org/10.1007/978-3-319-99395-9_1

One of the main goals of the ongoing research is focused on the optimization of proper electrolyte materials able to substitute first-generation electrolytes based on YSZ, along with the optimization of synthesis and sintering methods ready to be scalable to industrial level, and dopant/co-dopant compositions tuning their final electrochemical and microstructural properties. Among the proposed candidates, doped ceria-based systems as IT-SOFCs electrolytes appear to be the most relevant in terms of published results and of worldwide research interest.

Unfortunately, to the present-day ceria exhibits two major drawbacks leading to a barrier against its commercialization (even though several small companies using ceria-based systems as SOFC electrolytes already exist): not negligible electronic conductivity under SOFC operating conditions (due to partial reduction of Ce^{4+} to Ce^{3+}) and generally poor mechanical strength (due to sintering issues and bad quality of precursor powders). Considerable attempts have been made to overcome the former by using double-layered electrolytes, by introducing extra ultra-thin layers as a barrier for electrons, or by proper dopants/co-dopants addition. The present work has been mainly focused on the latter, by investigating the effects of different synthesis methods and/or sintering techniques with consequent optimization of several crucial parameters able to tune the produced electrolytes desired features.

After the first chapter providing an overview about fuel cell technology and, in particular, about SOFCs, Sects. 4.3 and 4.4 highlight the different microstructural and morphological features of both co-precipitate and hydrothermally synthesized doped ceria powders by analyzing different synthesis parameters and/or environments, clarifying both advantages and disadvantages of these particularly favorable processes pointed out by recent literature and accurately describing the engineered production process.

By considering Chap. 5, alternative sintering methods and their effects on ceria-based solid electrolytes are described. Fast firing (FF) and flash sintering (FS) have been proven to be successfully applied to different ceria-based systems. FF requires especially reactive nanopowders but with consequent drastic reduction of sintering temperatures and soaking times (i.e. from hours to minutes). Conversely, FS works for a wider range of powder compositions and synthesis conditions, still allowing a significant reduction of sintering temperature, paired in several cases with almost ideal microstructures.

Finally, one appendix provides a useful theoretical background about ion conduction and conventional sintering processes.

In conclusion, this work summarizes all the obtained results, by giving useful guidelines to produce dense ceria-based electrolytes for IT-SOFCs. In particular, an industrial-scale production should combine optimized powders with alternative sintering methods in order to achieve both ideal microstructure and substantial reduction of sintering temperature and time of a wide range of ceria-based IT-SOFC electrolytes.

Chapter 2
Fuel Cells

Fuel cells (FCs) are electrochemical devices able to directly convert fuels chemical energy into electrical energy with high efficiency and low environmental impact. Since the intermediate steps involving production of mechanical work by heat are avoided, FCs do not undergo thermodynamic limitations of heat engines (i.e. Carnot efficiency). Moreover, avoiding combustion leads to power production with minimal pollutants. However, unlike batteries, reductant and oxidant in fuel cells must be continuously replenished to grant continuous operation.

Even if FCs could process a wide variety of fuels and oxidants, the most interesting FCs are the ones that use common fuels and/or hydrogen as reductant and ambient air as oxidant.

A fuel cell core is constituted by the unit cells. The basic physical structure of a fuel cell consists of an electrolyte layer contacting an anode and a cathode on either side. Figure 2.1 gives a schematic representation of a unit cell.

In a typical FC, fuel is continuously fed to the anode (i.e. negative electrode) and an oxidant, generally atmospheric oxygen, is continuously fed to the cathode (i.e. positive electrode). The electrochemical reactions take place at the electrodes to produce an electric current flowing through the electrolyte, while driving an additional electric current making the load work. Although a fuel cell is similar to a common battery, it differs from it in many ways. In particular, a battery is an energy storage device where the available energy (or the reductant) is stored within itself, and a battery will cease to produce electrical energy when all the reactants are consumed (i.e. the battery is discharged). Conversely, a fuel cell is an energy conversion device where fuel and oxidant are continuously supplied, producing electrical energy as long as they are.

A critical portion of most unit cells is usually referred to as the three-phase interface. These mostly microscopic regions, in which the electrochemical reactions really take place, lie where either electrode meets the electrolyte [1]. For such areas to be reactive, they must be exposed to the reactant, be in electrical contact with the

L. Spiridigliozzi, *Doped-Ceria Electrolytes*, SpringerBriefs in Applied Sciences and Technology, https://doi.org/10.1007/978-3-319-99395-9_2

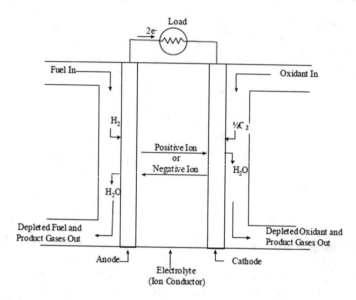

Fig. 2.1 Schematic representation of a FC unit cell

electrode, be in ionic contact with the electrolyte, and contain the appropriate electro-catalyst to let the reaction proceed at a desired rate. Density and nature of such three-phase interfaces play a crucial role in the electrochemical performance of both liquid and solid electrolyte FCs.

In liquid electrolyte FCs, the reactant gases diffuse through a thin electrolyte film wetting the porous electrode and electrochemically reacting on their related electrode surface. If the porous electrode contains an excessive amount of electrolyte, the electrode may flood and limit the gaseous species transport with consequent reduction in its performances. Hence, a delicate equilibrium must be maintained among the electrode, electrolyte, and gaseous species in the whole structure.

In solid electrolyte FCs, a great number of efficiently exposed to reactant gases cat-alyst sites that have to be electronically and ionically connected to electrode and elec-trolyte, respectively, must be assured. In most successful FCs, a high-performance interface requires electrodes having mixed ionic and electronic conductivity in areas near the catalyst.

Each of the unit cell components has to facilitate electrochemical reactions and also to conduct other critical functions. In particular, the electrolyte conducts ionic charge between the electrodes (completing the cell electric circuit shown in Fig. 2.1), also providing a physical barrier to prevent fuel and oxidant gas from a direct mix. Electrodes, on the other hand, conduct electrons away from the three-phase interface once they are formed (providing current collection and connection with the exter-nal load), ensure a homogeneous distribution of the reactant gases, and ensure that reaction products are efficiently led away to the bulk gas phase.

Over the past decades, unit cell performances have been dramatically improved (at least some of FC technologies), mainly resulted from three-phase boundary improvements, electrolyte thickness reduction, and operating temperature range broadening. Therefore, the ongoing research sees FCs as a promising technology for use as a source of heat and electricity for buildings, and as an electrical power source for electric motors propelling vehicles.

FCs can be classified according to either electrolyte or fuel choice, although the most common classification is by the electrolyte type. It includes polymer electrolyte fuel cells (PEFCs), alkaline fuel cells (AFCs), phosphoric acid fuel cells (PAFCs), molten carbonate fuel cells (MCFCs), and solid oxide fuel cells (SOFCs). Broadly, electrolyte choice governs the fuel cell operating temperature range, related in turn to thermomechanical properties of materials used for the other cell components [2].

Figure 2.2 shows an overview of the key features of the main fuel cell types, classified according to electrolyte choice.

Some FCs are also classified by the type of the used fuel. They include direct alcohol fuel cells (DAFCs), direct methanol fuel cells (DMFCs), and direct carbon fuel cells (DCFCs).

2.1 Polymer Electrolyte Fuel Cells (PEFCs)

A PEFC electrolyte is an ion exchange membrane (i.e. fluorinated sulfonic acid polymer or other similar polymers) being an excellent proton conductor. Generally, carbon electrodes with platinum as electro-catalyst are used for both anode and cathode, and with either carbon or metal interconnections.

The only liquid involved in PEFCs is water, thus limiting corrosion problems. However, water management in the membrane is crucial for efficient electrical performances. In fact, this fuel cell must operate under conditions involving water evaporation slower than its production because the polymer membrane must remain always hydrated. Hence, PEFCs operate at temperature ranging from 60 to 80 °C and under H_2-rich reactant gas with almost no CO, being it a FC poison at such low temperature. Moreover, high catalyst loadings (i.e. Platinum in most cases) are required for both anode and cathode. Finally, as mentioned above, the anode is easily poisoned by even traces of CO, sulfur species, and halogens (Fig. 2.3).

PEFCs are being developed for a wide range of applications and, in particular, for fuel cell vehicles (FCVs), the investment in PEFCs surpassing all other kinds of FCs combined.

The main advantages of PEFCs are the presence of a solid electrolyte providing excellent resistance to gas crossover, low operating temperature allowing quick device start-up, and the absence of any corrosive or "exotic" cell constituents.

	PEFC	AFC	PAFC	MCFC	SOFC
Electrolyte	Hydrated Polymeric Ion Exchange Membranes	Mobilized or Immobilized Potassium Hydroxide in asbestos matrix	Immobilized Liquid Phosphoric Acid in SiC	Immobilized Liquid Molten Carbonate in LiAlO$_2$	Perovskites (Ceramics)
Electrodes	Carbon	Transition metals	Carbon	Nickel and Nickel Oxide	Perovskite and perovskite / metal cermet
Catalyst	Platinum	Platinum	Platinum	Electrode material	Electrode material
Interconnect	Carbon or metal	Metal	Graphite	Stainless steel or Nickel	Nickel, ceramic, or steel
Operating Temperature	40 – 80 °C	65°C – 220 °C	205 °C	650 °C	600-1000 °C
Charge Carrier	H$^+$	OH$^-$	H$^+$	CO3$^=$	O$^=$
External Reformer for hydrocarbon fuels	Yes	Yes	Yes	No, for some fuels	No, for some fuels and cell designs
External shift conversion of CO to hydrogen	Yes, plus purification to remove trace CO	Yes, plus purification to remove CO and CO$_2$	Yes	No	No
Prime Cell Components	Carbon-based	Carbon-based	Graphite-based	Stainless-based	Ceramic
Product Water Management	Evaporative	Evaporative	Evaporative	Gaseous Product	Gaseous Product
Product Heat Management	Process Gas + Liquid Cooling Medium	Process Gas + Electrolyte Circulation	Process Gas + Liquid cooling medium or steam generation	Internal Reforming + Process Gas	Internal Reforming + Process Gas

Fig. 2.2 Schematic overview of the main FC types

The low and narrow operating temperature range, however, makes thermal management quite difficult, especially under high current density conditions. Moreover, water management represents another significant challenge in a PEFC design, as balance between sufficient membrane hydration against electrolyte flooding must be always assured.

Fig. 2.3 PEFC working principle

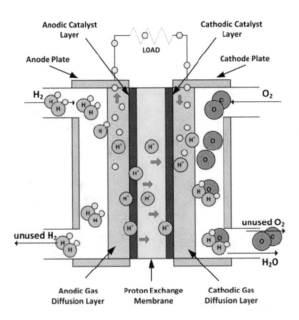

Even though some of these disadvantages can be avoided by lowering operating current density and/or increasing electrode catalyst loading, both interventions are quite expensive. Finally, hydrogen-fueled PEFCs need the development of hydrogen infrastructures, thus posing a barrier to extensive commercialization.

2.2 Alkaline Fuel Cells (AFCs)

An AFC electrolyte consists of concentrated (i.e. 85 wt%) KOH if the operating temperature is high (i.e. about 250 °C) or diluted (i.e. 35–50%) KOH if the operating temperatures are lower (i.e. <120 °C). The electrolyte is trapped in an asbestos matrix, and a wide range of electro-catalysts, such as Ni, Ag, metal oxides, noble metals, or spinels, can be used. Fuel supply is limited to non-reactive constituents, except for hydrogen. CO and CO_2 are poison for an AFC electrolyte, due to their ability to react with it by forming K_2CO_3, even in small amounts. Hence, hydrogen can be considered as the most preferable fuel for AFCs, although some direct carbon fuel cells use alkaline electrolytes (different from KOH) (Fig. 2.4).

AFCs have been firstly developed in the 1960s, especially to provide onboard electric power for spatial applications. They enjoyed considerable success thanks to their excellent performances on pure hydrogen and oxygen as reactant fuels and to their flexibility to effectively use a wide range of electro-catalysts.

However, the electrolyte sensitivity to CO_2 strictly requires the use of highly pure H_2 as fuel, thus requiring a highly effective reformer in terms of CO and CO_2 removal.

Fig. 2.4 AFC working principle

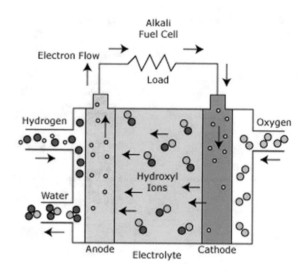

Moreover, if ambient air is used as the oxidant, ambient CO_2 must be also removed. While technologically not challenging, ambient CO_2 removal has a significant impact on size and cost of the whole system. In conclusion, although AFCs are still used for spatial applications, their terrestrial applications are almost negligible.

2.3 Phosphoric Acid Fuel Cells (PAFCs)

Phosphoric acid (100% concentrated), retained in a silicon carbide matrix, is used as electrolyte for PAFCs typically operating at temperatures ranging from 150 to 220 °C. At lower temperatures, phosphoric acid is a poor ionic conductor and CO poisoning of the Pt-based electro-catalyst in the anode becomes severe. Phosphoric acid is highly stable compared to other common acids, making PAFCs capable of operating at the high end of acid temperature range (i.e. 100–220 °C). Moreover, the use of concentrated acid minimizes water vapor pressure, leading to easy water management (Fig. 2.5).

PAFCs are mostly developed for stationary applications, as many different PAFC-based systems are commercially available. However, PAFCs development slowed down over the last few years, in favor of PEFCs, although PAFCs are much less sensitive to CO than PEFCs and AFCs (tolerating up to one percent of CO as diluent). Another PAFC advantage consists in considerable design flexibility for thermal management, due to a relatively wide operating temperature range. In addition, waste heat from PAFCs operations can be readily used in most commercial and industrial cogeneration applications, technically allowing the use of a bottoming cycle.

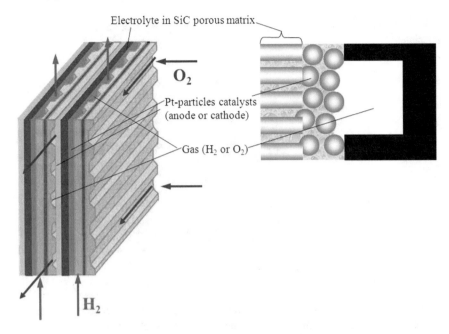

Electrolyte in SiC porous matrix

O_2

Pt-particles catalysts (anode or cathode)

Gas (H_2 or O_2)

H_2

Fig. 2.5 PAFC working principle

PAFCs main disadvantages consist in a slow cathode-side oxygen reduction requiring the use of a platinum catalyst, in an often required extensive fuel processing including a prior efficient water gas shift reactor and in the highly corrosive nature of phosphoric acid itself requiring the use of expensive stack materials.

2.4   Molten Carbonate Fuel Cells (MCFCs)

A MCFC electrolyte is usually a combination of alkali carbonates retained in a ceramic matrix of $LiAlO_2$. These fuel cells operate at temperatures generally ranging from 600 to 700 °C, where the alkali carbonates form a highly conductive molten salt, the carbonate ions providing ionic conduction. Thanks to the high operating temperatures in MCFCs, Ni (in the anode) and nickel oxide (in the cathode) can be used as electro-catalysts to promote reaction, as noble metals are not strictly required and many common hydrocarbon fuels can be internally reformed.

MCFCs development focus has mainly been stationary and marine applications, as the relatively large MCFCs size and weight and their slow start-up time are not an issue for such systems. MCFCs are also under development for use with a wide range of conventional and renewable fuels, and MCFC-like technology is even considered for direct carbon fuel cells (DCFCs) (Fig. 2.6).

Fig. 2.6 MCFC working
principle

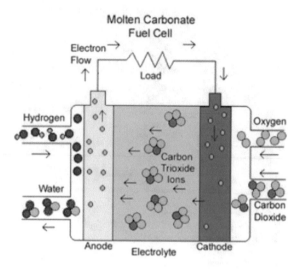

The relatively high MCFCs operating temperatures result in several benefits, including no need of expensive electro-catalysts as the nickel-based electrodes provide adequate activity and possibility to use CO and certain hydrocarbons as fuels as they are converted to hydrogen on special reformer plates within the stack. In addition, as for PAFCs, waste heat can be used for a bottoming cycle further increasing the system efficiency. Conversely, high temperatures promote material issues, thus impairing mechanical stability and stack life.

However, the main challenge in MCFCs development is represented by the very corrosive and mobile electrolyte, which requires the use of both nickel and high-grade stainless steel as the cell "hardware." In addition, a source of CO_2 (usually recycled from anode exhaust) is always required at the cathode to form new carbonate ions, representing an additional system complexity.

2.5 Solid Oxide Fuel Cells (SOFCs)

A SOFC electrolyte is a solid, nonporous metal oxide operating at temperatures ranging from 600 to 1000 °C, where ionic conduction by oxygen ions takes place. Typically, the anode consists of a $Ni-ZrO_2$ cermet and the cathode consists of Sr-doped $LaMnO_3$.

SOFCs are the fuel cells with the longest continuous development period, approximately starting in the late 1950s, and early on, the limited conductivity of solid electrolytes required cell operations at around 1000 °C. As the time went on, thinner or different electrolytes along with improved cathodes allowed a reduction in operating temperature down to 650–850 °C, even if the ongoing research is attempting to push SOFCs operating temperatures even lower (i.e. 500–650 °C) in the so-called intermediate-temperature SOFCs (IT-SOFCs) (Fig. 2.7).

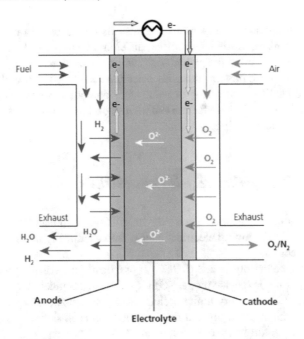

Fig. 2.7 SOFC working principle

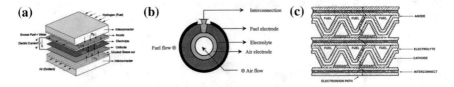

Fig. 2.8 Different SOFC stack configurations: **a** planar, **b** tubular, and **c** monolithic

Because of the various performance improvements, SOFCs are now considered for a wide range of applications, including stationary power generation, mobile power, auxiliary power for vehicles, and specialty applications.

Since SOFCs electrolyte is solid, the cell can be cast into various shapes, leading to consequent stack configurations such as tubular, planar, or monolithic. Moreover, the solid ceramic construction of the unit cell reduces any corrosion problems and allows a precise engineering of the three-phase boundary, by avoiding electrolyte movement or any flooding in the electrodes. SOFCs kinetics are relatively fast, CO is a directly useable fuel as in MCFCs, and there is no particular requirement for CO_2 at the cathode. Another SOFC advantage is represented by the modest cost of the required materials, being usually no need of expensive electro-catalysts. Finally, SOFCs operating temperatures allow using most of the waste heat for cogeneration or bottoming cycles, with overall system efficiencies ranging from 40 up to 60% (Fig. 2.8).

SOFCs high operating temperature has also its drawbacks. In particular, there are thermal expansion mismatches among different materials, and sealing between unit cells can be difficult in the planar configuration. Additionally, high temperature imposes severe constraints on materials selection resulting in difficult fabrication processes and facilitates corrosion of metal stack components such as some design interconnects. Definitely, these factors can limit stack-level power density along with thermal cycling and stack life.

2.5.1   Solid Oxide Electrolysis Cells (SOECs)

An electrolysis cell consumes electricity to split water molecules into hydrogen and oxygen, by transforming electrical energy into chemically bound energy in the hydrogen molecules.

A solid oxide electrolysis cell (SOEC) is practically the corresponding fuel cell (i.e. SOFC) running in reverse mode. SOECs operate at relatively high temperatures (i.e. 700–1000 °C) with very high efficiencies. Hydrogen and oxygen, i.e. the two electrolysis products, are formed at the electrodes on each side of the cell. SOECs could be used for hydrogen production from surplus electricity generated by renewable systems such as wind turbines or solar cells. Then, the hydrogen can be stored and reconverted into electricity (by using fuel cells in SOFCs mode) if demand arises (Fig. 2.9).

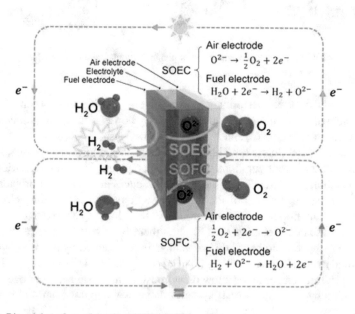

Fig. 2.9 Bimodal configuration of a SOFC/SOEC-based system

SOECs can also electrolyze carbon dioxide to carbon monoxide, and if water is electrolyzed at the same time (co-electrolysis mode), it is possible to produce a gaseous mixture of hydrogen and carbon monoxide. Such mixture, comparable to a syngas, is the starting point of a great number of hydrocarbon synthesis in the chemical industry. Hence, liquid transport fuels can be synthetically produced in this way. Moreover, if the electricity is produced by renewable sources, the use of such fuel is CO_2 neutral.

References

1. EG&G Technical Services, Inc., *Fuel Cell Handbook* (U.S. Department of Energy, Morgantown, 2004)
2. R. O'Hayre, S. Cha, W. Colella, F.B. Prinz, *Fuel Cell Fundamentals* (Wiley, Hoboken, 2016)

Chapter 3
SOFC Components

SOFCs major components (as for the other fuel cells) are the anode, the cathode, and the electrolyte. Fuel cell stacks contain also an electrical interconnect linking individual unit cells together in series or parallel.

The electrolyte is made from ceramic conducting oxide ions. Oxygen atoms are reduced into oxide ions on the porous cathode surface by electrons and then flow through the solid electrolyte to the fuel-rich porous anode where the oxygen ions can react with the used fuel by giving up electrons. Finally, the interconnection lets the electrons flow through an external load circuit.

SOFCs electrolytes are not sensitive to membrane hydration (as in PEFCs) and do not require water management systems. However, in absolute terms, ionic conductivity in ceramic oxide is well below that of most polymeric proton conductors. Hence, to have sufficiently high ionic conduction through such oxide membranes, SOFCs operating temperatures have to be reached. There are many candidates as SOFCs electrolyte materials, most notably fluorite-structured materials such as zirconia-based and ceria-based systems and some doped perovskite systems.

SOFCs electrode materials must simultaneously provide high porosity, high electrical conductivity, and high catalytic activity. However, common conductive materials such as carbon and most metals are unsuitable for SOFC high-temperature environments. Furthermore, electrode materials have to face severe thermal cycles during SOFC start-up and shut down with consequent mechanical stability issues. Hence, most SOFC electrode materials are either electrically conductive ceramics or mixed ceramic–metal composites (i.e. cermets) in order to ensure thermal and chemical compatibility under severe operating conditions.

© The Author(s), under exclusive license to Springer Nature Switzerland AG 2018
L. Spiridigliozzi, *Doped-Ceria Electrolytes*, SpringerBriefs in Applied Sciences
and Technology, https://doi.org/10.1007/978-3-319-99395-9_3

3.1 Anode Materials

A wide range of materials has been potentially considered as anode materials for SOFCs.

Currently, most SOFCs anodes are made of Ni–YSZ cermets, as they meet most of the electrode requirements discussed in this chapter. A Ni–YSZ cermet anode is typically prepared by sintering NiO and YSZ powders via solid-state synthesis. The resulting oxide composite is then reduced upon exposure to fuel gases, transforming its final structure. In such anodes, metallic nickel provides electronic conductivity and catalytic activity, whereas YSZ provides structural stability, thermal expansion matching, and ionic conductivity to the electrode, thus effectively widening the three-phase interfaces. Ni and YSZ are definitely inert and immiscible into each other over a wide temperature range, being chemically stable in reducing environments even at high temperatures. Moreover, charge transfer resistance associated with electro-chemical reaction at the Ni–YSZ boundary is low, thus ensuring good performances as electro-catalyst. However, the major advantage of Ni–YSZ cermets is the ability to closely match its thermal expansion coefficient to that of the electrolyte (especially if the electrolyte is a zirconia-based system) by tuning Ni:YSZ ratio, Ni:YSZ particle ratio, and the amount of present porosity.

Ni–YSZ anodes present several disadvantages too, including the tendency to show performance degradation upon prolonged operation caused by Ni coarsening, agglomeration and/or oxidation, the low tolerance to sulfur impurities of the fuel gas, and the tendency to form carbon deposits under hydrocarbon fuels operating conditions.

Recently, interest in doped ceria-based materials as SOFC anodes has arisen. As a primary advantage of their use as anode materials, there is their ability to overcome carbon deposition issues, subsequently facilitating the use of hydrocarbons as SOFCs fuels. Doped ceria (like metallic nickel) is a good electro-catalyst for methane oxidation, and being both electronic and ionic conductor under reducing conditions, an electrochemical reaction can directly proceed on a doped ceria anode.

To use ceria in SOFC anodes, however, it is mandatory to solve the mechanical issues related to ceria partial reduction from Ce^{4+} to Ce^{3+}. In fact, this transition results in a lattice expansion, possibly causing mechanical failure due to electrode –electrolyte cracking and delamination. Doping with higher concentration (i.e. about 40 mol%) of lower valent cations than the usually used can potentially increase the dimensional stability of the anode. Moreover, ceria-based anodes performance can be significantly improved by adding Ni or noble metals to them, improving also their catalytic activity toward methane oxidation. Finally, ternary anodes composed of doped ceria, stabilized zirconia, and metallic Cu have been also investigated, being extremely promising material system for SOFC anodes.

A wide variety of perovskite oxides have also been considered as potential candidates for SOFC anodes. As for ceria-based systems, the main advantage of perovskite-based anodes is their ability to suppress carbon deposition, thus allowing a direct use of hydrocarbon fuels. Candidate perovskite anodes that can be found in the literature include lanthanum chromite-based systems, oxygen-deficient doped per-ovskite, and LSCV (i.e. $La_{0.8}Sr_{0.2}Cr_{0.97}V_{0.03}O_3$)—YSZ composites. Pure lanthanum chromite is not a good SOFC anode material due to problematic lattice expansion and p-type electrical conductivity, but doping it with Sr and Ti leads to sufficiently good anode materials. However, overall doped lanthanum chromite performances are quite uncompetitive compared to Ni–YSZ cermets.

In addition to anode material candidates discussed above, a variety of other poten-tial anode materials being recently researched to further optimize SOFC perfor-mances include tungsten bronze oxides and pyrochlore-type oxides.

3.2 Cathode Materials

SOFCs cathodes have to provide high activity for the electrochemical reduction of oxygen as well as both ionic and electronic conductivity to maximize the num-ber of triple-phase boundary sites where the reduction can occur. These conditions are achieved by using composite cathodes and/or mixed ionic–electronic conduc-tor (MIEC) materials. Because metal conductors are generally not stable in high-temperature oxidizing environments, SOFC cathodes are usually pure ceramics. Hence, electronic conductivity in SOFC cathodes is generally much lower than in SOFC anodes, where metal–ceramic cermets can be used. Such restriction leads to optimization issues toward cathode thickness. In fact, ionic transport primarily occurs by ions flowing in a direction normal to cathode surface (resulting in ionic conductivity being directly proportional to cathode thickness), whereas electronic transport primarily occurs by electrons flowing in a direction parallel to cathode surface (resulting in electronic conductivity being inversely proportional to cathode thickness). Therefore, the optimal cathode thickness must be derived by minimizing the total ionic and electronic resistance.

In YSZ-based SOFCs, the dominant cathode material is a strontium-doped $LaMnO_3$ perovskite (LSM), because of its good physical/chemical stability, elec-trical conductivity, and catalytic activity. Unfortunately, LSM has a very low oxygen ion conductivity, and therefore, LSM-based cathodes are usually mixed with YSZ to form LSM–YSZ composite cathodes, YSZ providing the required ionic conductivity.

MIEC alternatives to LSM–YSZ cathodes are currently under development. In particular, Fe-doped lanthanum cobaltite (LSCF) as potential IT-SOFCs cathode is particularly promising, the iron incorporation being crucial for preventing possible reactions with the electrolyte.

3.3   Electrolyte Materials

Oxide ions conducting solid electrolytes cover a wide range of ceramic materials, which mainly include fluorite-/perovskite-/brownmillerite-structured materials.

The cubic fluorite structure is, basically, the most common electrolyte structure, being the structure of both ceria and doped zirconia. More in detail, ceria is almost always stable in the fluorite form while zirconia can be stabilized in its cubic structure at low temperatures by doping it with divalent and/or trivalent cations, typically with 8 mol% Y^{3+} (8YSZ) [1]. Although ceria is stable in its fluorite form even at room temperature, a doping with divalent and/or trivalent cations is always needed in order to increase oxygen vacancies and, consequently, its ionic conductivity. Gd^{3+} and Sm^{3+} are commonly used dopants for ceria-based electrolytes [2]. Another fluorite-structured electrolyte material that has been proposed is the δ-Bi_2O_3: a polymorphic form of bismuth oxide owning an oxygen-deficient fluorite structure with a quarter of the normal anion sites vacant [3]. However, despite its very high ionic conductivity, δ-Bi_2O_3 presents serious drawbacks in terms of chemical instability at low temperatures. In fact, at room temperatures, it strives to assume its monoclinic polymorphic form and it is easily reduced at low oxygen partial pressures, decomposing itself into metallic bismuth and oxygen [4]. Therefore, a real usage of bismuth oxide as SOFCs electrolytes is practically negligible.

3.3.1   Zirconia-Based Materials

Zirconia stabilized in the fluorite form is the most commonly used material as SOFCs electrolyte, thanks to its adequate ionic conductivity, its chemical stability, and its mechanical properties. Doping with rare earths, alkaline earths, and lanthanide oxides stabilizes monoclinic zirconia to cubic zirconia over a wide range of temperatures, simultaneously enhancing its ionic conductivity to an extended oxygen partial pressure range of SOFC operations. Yttria (Y_2O_3), calcia (CaO), magnesia (MgO), scandia (Sc_2O_3), ytterbia (Yb_2O_3), and dysprosia (Dy_2O_3) have all been used as dopants to stabilize zirconia in its fluorite form, even though yttria is widely and commercially used because of its abundance and its cost-effectiveness compared to the other oxides. Yttria (8 mol%) resulted to be the most effective dopant composition in terms of conductivity enhancing and fluorite structure stabilizing. Typical conductivity values of YSZ at high temperatures (800–1000 °C) are in the order of magnitude of $10^{-1}/10^{-2}$ S/cm. Scandia-doped zirconia (ScSZ) has also found great interest in the literature [5, 6] thanks to its much higher conductivity than YSZ, even though the high cost of scandium and negative aging effects of ScSZ electrolytes impair their real commercialization in SOFCs (Fig. 3.1).

Fig. 3.1 Fluorite-like YSZ
crystal structure

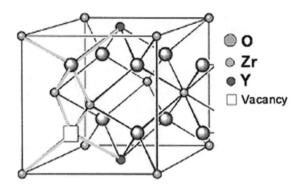

3.3.2 Perovskite-Structured Materials

According to the literature, also some doped perovskite materials (ABO_3) can be used as electrolytes for SOFCs. Among them, Al- or Mg-doped calcium titanate ($CaTiO_3$) has one of the highest ionic conductivities [7], but still much lower than YSZ and doped ceria conductivities. Another perovskite-structured material whose properties have been investigated in the literature is lanthanum strontium gallium magnesium oxide ($La_{0.9}Sr_{0.1}Ga_{0.8}Mg_{0.2}O_3$) [8]. Although it exhibits ionic conductivity higher than zirconia-based and ceria-based electrolytes, its very poor mechanical properties make it unsuitable for real SOFCs applications (Fig. 3.2).

Fig. 3.2 $SrTiO_3$ perovskite
crystal structure

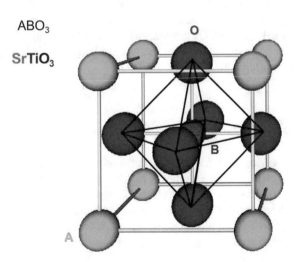

Fig. 3.3 $Sr_2Fe_2O_5$
brownmillerite crystal
structure

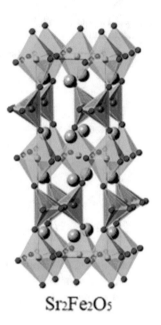

$Sr_2Fe_2O_5$

3.3.3   Brownmillerite-Structured Materials

A brownmillerite-structured compound has a general formula like $A_2B_2O_5$: a perovskite-structured material with a sixth of oxygen sites vacant. A typical example that can be found in the literature of this kind of SOFCs possible materials is $Ba_2Ln_2O_5$ [9]. Owning an ionic conductivity higher than YSZ with a large increase of it at about 900 °C, this barium lanthanate could be used as electrolyte material for high-temperature SOFCs (Fig. 3.3).

3.3.4   LAMOX Materials

$La_2Mo_2O_9$ (LAMOX) is a relatively newly discovered fast oxygen ion conductor, with huge conductivities at high temperatures compared to zirconia-based materials [10]. However, undoped LAMOX typically undergoes to a reversible phase transition at around 600 °C from a nonconductive monoclinic form to the very highly conductive fluorite form; this phase transition, associated also with possible mechanical breakdown after repeated cycles, is the main drawback of such material. Praseodymium-doped LAMOX appears to lower down the temperature of that phase transition, but it does not completely suppress it [11]. Finally, another drawback of LAMOX materials is the proneness of molybdenum to reduction at low oxygen pressures, making them also unsuitable for commercialization as SOFCs electrolytes (Fig. 3.4).

Fig. 3.4 La$_2$Mo$_2$O$_9$ crystal structure

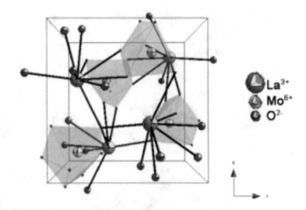

3.3.5 Ceria-Based Materials

Ceria-based materials are the most promising candidate as electrolyte materials for IT-SOFCs [12–14]. Thanks to a better ionic conductivity in the operative IT-SOFCs temperature range compared to zirconia-based materials, doped ceria electrolytes attracted enumerable studies all over the world. In particular, many efforts have been made to establish the best dopant and dopant concentration for ceria-based systems, but now the literature identifies those dopants in Gd^{3+} and Sm^{3+} and their optimal composition as 10 and 20%, respectively [15, 16]. However, ceria-based electrolytes present a major drawback in terms of not negligible electronic conduction at low oxygen partial pressure and of potential reduction and mechanical destruction under great oxygen partial pressure gradients typical of SOFCs operating conditions [17]. Pure cerium oxide can be either hexagonal Ce$_2$O$_3$ or cubic CeO$_2$, with several intermediate compositions depending on chemical and physical conditions. Its color is a strong yellow when cerium is Ce^{4+} while becoming violet blue when it is reduced to Ce^{3+}. Cerium oxide color can also be influenced by the presence of different lanthanide dopants; in particular, its color can converge to dark yellow and up to brown (Fig. 3.5).

Stoichiometric ceria is not a good electrolytic material due to its low ionic and electronic conductivity, but its easiness of reducing itself into non-stoichiometric conditions after mild heating treatments (500–700 °C) makes cerium oxide a very interesting material. Moreover, ceria (CeO$_2$) structure allows a wide number of dopant cations into its crystal lattice (such as rare-earths or alkali-earths cations), thus achieving very good catalytic and electrochemical properties, potentially becoming either an electronic, ionic, or a mixed ionic–electronic conductor [18].

Rare-earths or alkali-earths oxides usually are highly soluble into ceria, up to 30–40% depending on different elements, easily forming stable solid solutions characterized by different properties compared to the undoped material. Among the large number of doping oxides that can be found in the literature, the most commonly used are Ca, Mg, Ba, Sr, Pr, Y, Gd, and Sm oxides [16].

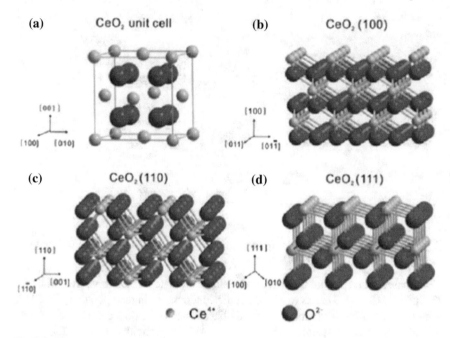

Fig. 3.5 Atomic configuration of CeO₂ unit cell (**a**) and its {100} (**b**), {110} (**c**), {111} (**d**) facets

Generally speaking, highly ionic conducting oxides own an open crystalline structure (like the fluorite cubic structure) and exhibit good tolerance to atomic disorder introduced by doping and co-doping and to changes in oxidation number. Ceria, almost always stable in the fluorite structure, is one of the best examples of such conducting oxides, and by introducing aliovalent cations into its cubic structure, it is possible to create oxygen vacancies that are the mainly responsible of oxygen ions conduction. In particular, ionic conductivity can be expressed by the following equation:

$$\sigma T = \sigma_0 \exp(-Ea/kT) \tag{3.1}$$

where σ_0 is the pre-exponential factor (independent from dopant concentration at low temperatures), E_a is the conduction activation energy, k is the Boltzmann constant, and T is the temperature in K. The activation energy can be expressed in turn as $\Delta H_m +$ ΔH_a, where ΔH_m is the vacancy/cation migration enthalpy and ΔH_a is the vacancy association enthalpy. In order to optimize free oxygen vacancies concentration, ΔH_a contribute has to be minimized [19].

Association enthalpy, depending on the electrostatic interactions generated by vacancies charge inside the lattice, can be related to the doping cation ionic radius. In particular, when the ionic radius of the doping cation is similar to the one of hosting cation, association enthalpy reaches a minimum and the resulting ionic conductivity of the doped material is maximized. Based on these considerations, the best doping cations for ceria are, indeed, Gd and Sm. Optimal values of doping concentrations, however, are different for each cation, being the whole conduction activation energy dependent on absolute temperature. Hence, by considering the operating temperatures of IT-SOFCs electrolyte, the optimal compositions for gadolinium-doped ceria (GDC) and samarium-doped ceria (SDC) have been found to be 10 mol% [19] and 20% [16], respectively.

References

1. R. O'Hayre, S. Cha, W. Colella, F.B. Prinz, *Fuel Cell Fundamentals* (Wiley, Hoboken, 2016)
2. N.P. Bansal, P. Singh, *Advances in Solid Oxide Fuel Cells V* (Wiley, Hoboken, 2010)
3. T. Takahashi, H. Iwaharae, Y. Nagai, High oxide ion conduction in sintered Bi_2O_3 containing SrO, CaO or La_2O_3. Journal of Applied Electrochemistry **2**(2), 97–104 (1972)
4. M.J. Verkerk, A.J. Burggraaf, High oxygen ion conduction in sintered oxides of the Bi_2O_3–Ln_2O_3 system. Solid State Ionics, 463–467 (1981)
5. G. Accardo, G. Dell'Agli, D. Frattini, L. Spiridigliozzi, S.W. Nam, S.P. Yoon, Electrical behaviour and microstructural characterization of magnesia co-doped ScSZ nanopowders synthesized by urea co-precipitation. Chem. Eng. Trans. **57** (2017)
6. H.J. Choi, M. Kim, K.C. Neoh, H.J. Kim, J.M. Shin, G. Kim, J.H. Shim, High-performance silver cathode surface treated with scandia-stabilized zirconia nanoparticles for intermediate temperature solid oxide fuel cells. Adv. Energy Mater. (2016)
7. M. Takahashi, Space charge effect in lead zirconate titanate ceramics caused by the addition of impurities. Japan. J. Appl. Phys. **9**(10) (1970)
8. Z. Naiqing, S. Kening, Z. Deruie, J. Dechang, Study on properties of LSGM electrolyte made by tape casting method and applications in SOFC. J. Rare Earths **24**(1), 90–92 (2006)
9. S.C. Singhal, Advances in solid oxide fuel cell technology. Solid State Ionics **135**(1–4), 305–313 (2000)
10. F. Goutenoire, O. Isnard, E. Suard, O. Bohnke, Y. Laligant, R. Retouxe, P. Lacorre, Structural and transport characteristics of the LAMOX family of fast oxide-ion conductors, based on lanthanum molybdenum oxide $La_2Mo_2O_9$. J. Mater. Chem. **11**, 119–124 (2001)
11. D. Marreo-Lopez, J. Canales-Vazquez, J.C. Ruiz-Morales, J.T.S. Irvinee, P. Nunez, Electrical conductivity and redox stability of $La_2Mo_{2-x}Wx\ O_9$ materials. Electrochim. Acta **50**(22), 4385–4395 (2005)
12. C. Veranitisagul, A. Kaewvilai, W. Wattanathana, N. Koonsaeng, E. Traversae, A. Laobuthee, Electrolyte materials for solid oxide fuel cells derived from metal. Ceram. Int. **38**, 2403–2409 (2012)
13. F.F. Munoz, A.G. Leyva, R.T. Bakere, R.O. Fuentes, Effect of preparation method on the properties of nanostructured gadolinia-doped ceria materials for IT-SOFCs. Int. J. Hydrogen Energy **37**, 14854–14863 (2012)
14. S. Wang, C. Yeh, Y. Wange, Y. Wu, Characterization of samarium-doped ceria powders. J. Mater. Res. Technol. **2**(2), 141–148 (2013)
15. J. Stojmenovic, M. Zunic, J. Gulicovski, D. Bajuk-Bogdanovic, V. Dodevskie, S. Mentus, Structural, morphological, and electrical properties of doped ceria as a solid electrolyte for intermediate-temperature solid oxide fuel cells. J. Mater. Sci. **50**, 3781–3794 (2015)
16. G. Donmez, V. Saraboga, T. Gurkaynake, M.A.F. Oksuzomer, Polyol Synthesis and Investigation of $Ce_{1-x}RE_xO_{2-x/2}$ (RE = Sm, Gd, Nd, La, $0 \leq x \leq 0.25$) Electrolytes for IT-SOFCs. J. Am. Ceram. Soc. **2**, 501–509 (2015)

17. A. Selcuke, A. Atkinson, Elastic properties of ceramic oxides used in solid oxide fuel cells (SOFC). J. Eur. Ceram. Soc. **17**(12), 1523–1532 (1997)
18. A.J. Jacobson, Materials for solid oxide fuel cells. Chem. Mater. **22**, 660–674 (2010)
19. B.C.H. Steele, Appraisal of $Ce_{1-y}Gd_yO_{2-y/2}$ electrolytes for IT-SOFC operation at 500 °C. Solid State Ionics **129**(1–4), 95–110 (2000)

Chapter 4
Doped Ceria Electrolytes: Synthesis Methods

4.1 Solid-State Synthesis

The term solid-state synthesis is generally used to describe the interactions among solid reagents, where neither a solvent medium nor controlled vapor-phase interactions are utilized. Pure and well-characterized solid precursors are crucial for every solid-state reaction.

A solid-state synthesis (also called ceramic method) consists of heating of two or more nonvolatile solids that react to form the final desired product. This method can be used to synthesize an extremely large range of solid compounds, such as mixed metal oxides, sulfides, nitrides, aluminosilicates. A simple example of a solid-state reaction is the synthesis of zirconium silicate through the following reaction:

$$ZrO_2(s) + SiO_2(s) \xrightarrow{T=1300\,°C} ZrSiO_4(s) \tag{4.1}$$

High or very high temperature are generally required (600–2000 °C) for solid-state synthesis, because a significant amount of energy is needed to overcome lattice energies in order to let an ion diffuse into a different occupation site. In fact, solid-state reactions are diffusion controlled according to Fick's first law:

$$J = -D\frac{d\varphi}{dx} \tag{4.2}$$

where J is the diffusion flux, D is the diffusion coefficient (i.e. diffusivity), φ is the species concentration, and x is the position (in length). To maximize the diffusion flux and, in turn, to decrease reaction times (that may range from hours to several days or even weeks), it is possible to increase reaction temperatures (because of the Arrhenius dependence between D and the absolute temperature in solids) or to maximize surface contact area by achieving the smallest possible particle size of the reactants.

© The Author(s), under exclusive license to Springer Nature Switzerland AG 2018
L. Spiridigliozzi, *Doped-Ceria Electrolytes*, SpringerBriefs in Applied Sciences and Technology, https://doi.org/10.1007/978-3-319-99395-9_4

About reaction temperatures, the so-called Tamman's rule suggests to reach an absolute temperature of about two-thirds of the lower melting reactant's melting point in order to make the solid-state reaction occur in reasonable times.

About reactants particle size, a thorough grinding or a pelletizing step by using a hydraulic press are usually necessary to achieve homogeneous mixtures of reactants. Moreover, the reaction mixture is typically removed from crucibles and reground to bring fresh surfaces and ready to react in contact and further speeding up the whole reaction.

Finally, due to the high temperatures reached during solid-state synthesis, crucibles must be sufficiently inert to the reactants and withstand such temperatures without any degradation.

4.2   Wet Chemical Synthesis

Wet chemical synthesis is a group of synthesis technique carried on mostly in the liquid phase that allows the production of inorganic particles, either oxides or metals, possibly multicomponent phases, with controlled size and shape, high chemical/physical reactivity, and very high purity control. Such control on the obtained materials requires the development of synthesis usually starting from precursors in the form of metal salts aqueous solutions. These kinds of processes are based on the hydrolysis (in case of oxides) or on the reduction (in case of metals) of precursors dissolved in water or in organic solvents caused by pH changes (generally alkalization) or by the addition of reducing agents.

Wet chemical synthesis techniques generally used to produce nanometric or submicrometric powders of oxides which can be classified as:

- Water precipitation/co-precipitation;
- Hydrothermal synthesis;
- Solgel synthesis;
- Microwave-assisted synthesis.

The optimization of the produced materials is performed throughout an accurate tuning of all synthesis parameters, whose effects can easily modify the powders' desired properties. Depending on the synthesis method and by the post-synthesis treatments, a wide range of specific equipment has to be used: traditional heating systems, microwave ovens, refluxing systems, magnetic stirrers, temperature controllers, autoclaves and vessels for the hydrothermal environment, and so on.

4.3 Precipitation/Co-precipitation

Precipitation consists in the formation of a solid product starting from a solution. Co-precipitation is the simultaneous precipitation of different solutes. The solid formed after precipitation/co-precipitation usually is a low soluble compound, such as oxides, hydroxides, carbonates, or oxalates, and it is called *precipitate/co-precipitate*. The precipitate-free liquid is called *supernatant*, while the chemical agent that causes precipitation is called *precipitant* or *precipitating agent*.

Precipitation is based on pH modification (generally by raising it up) of a starting solution containing one or more metallic cations: Under these conditions, they begin to precipitate in solid form, separating from the starting solution and nucleating in the form of low or very low soluble compounds. The pH modification is obtained by adding to the solution a precipitating agent able to alkalize it and, consequently, make cations exceed their solubility and precipitate. The most common precipitating agents are ammonia, alkali hydroxides, alkali carbonates, organic bases, and urea. Based on the used precipitating agent (and on the anions brought to the solution), the formed precipitate can either be a carbonate, a hydroxide, a hydrated oxide, an oxalate, etc.

Precipitation may also occur if a compound's concentration exceeds its solubility due to temperature or pressure changing, i.e. supersaturating the starting solution.

After precipitate formation and its sedimentation, a filtration step is always needed to separate it from the remaining supernatant. Moreover, the obtained precipitate after filtration has to be accurately washed with either deionized water and/or alcohols in order to remove any undesired impurities. Finally, the product undergoes a low-temperature drying step to remove the adsorbed water and to an intermediate-temperature calcination step to obtain the final products.

Parameters influencing precipitation/co-precipitation are definitely manifolds. Among them, concentration of different involved components, temperature, pH of the starting solution, drying and calcination conditions, precipitating agent nature, mixing speed are the most important parameters to be controlled in order to obtain a final product with the desired characteristics in terms of particle size, agglomeration, morphology, and so on. Furthermore, precipitation can be classified as direct precipitation or reverse precipitation, depending on the mixing order between starting solution and precipitating agent. In the former case, the precipitating agent is directly added to the solution containing cations, while in the latter case, the starting solution is added to another solution containing the precipitating agent. Generally, reverse precipitation is the preferred configuration. In fact, direct precipitation can easily give rise to uneven growth of precipitate particles, thus impairing the quality of the final product.

The main advantages of precipitation/co-precipitation synthesis are low cost, great simplicity, low reaction temperature, uniform and fine/ultra-fine particle size of the obtained products. However, being this kind of synthesis strongly susceptible to many conditions, several possible drawbacks have to be faced: incomplete precipitation of metallic cations, uncontrolled particles morphology, agglomeration phenomena, wide distribution of particle size, and rather poor control over the stoichiometry of certain systems.

4.3.1 Urea-Based Homogeneous Precipitation (UBHP)

Basically, UBHP method is a precipitation/co-precipitation synthesis carried out by using urea ($CO(NH_2)_2$) as precipitating agent at its decomposing temperature. The following steps are involved in a typical UBHP: dissolution of precursors into an aqueous solution, dissolution of the required amount of urea into the same solution, heating at temperatures higher than ≈ 83 °C, homogeneous urea decomposition and consequent pH increase, formation of the precipitate [1] (Fig. 4.1).

At $T > 83$ °C, urea started to decompose according to the following global reaction:

$$CO(NH_2)_2 + H_2O \rightarrow CO_2 + 2NH_3 \qquad (4.3)$$

It can be proved that dissolved urea exists only as mono-molecules in aqueous solutions [2], and neither urea bi-molecules nor more conjugated molecules can exist [3]. The in situ urea decomposition leads to the active and homogeneous release of OH^- and CO_3^{2-} ions and, consequently, to solution alkalization and co-precipitate formation, thus avoiding undesired and localized distribution of the reactants. Thanks to these phenomena, nucleation and growth of the as-precipitate morphology can be easily controlled, compared to other precipitation syntheses. In fact, upon UBHP, precipitating ligands release can be easily controlled through temperature variation with consequent explosive nucleation at temperatures higher than urea decomposing temperature (i.e. 83 °C). More in detail, by controlling the heating rate of a solution containing urea to a set temperature, the precipitating ligands concentration can be

Fig. 4.1 pH evolution upon UBHP process

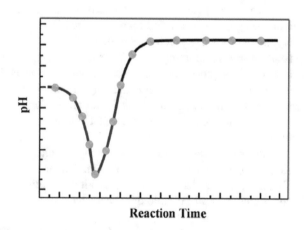

Reaction Time

tuned; i.e. by varying the cationic concentration, the total amount of growth nuclei can be easily controlled. According to the Arrhenius plot referring to urea decomposition in water, the calculated activation energy for this decomposition process results in 32.4 kcal [2].

Primary particles of the homogeneously formed precipitate consequently tend to aggregate to form larger particles maintaining their same features in terms of narrow size distributions and well-defined morphologies [4, 5]. The UBHP peculiar advantages have been exploited to synthesize a large variety of inorganic particles [3, 6, 7] as well as ceria-based systems [1, 8] (Fig. 4.2).

The co-precipitated ceria-based powders synthesized by UBHP are constituted by cerium carbonates, and the desired fluorite-structured oxide is still not formed. Usually, such cerium carbonate-based compounds exhibited a pale white color [1]. Diffraction analysis carried out on such as-precipitated powders showed that a crystalline oxycarbonate phase was formed, corresponding to cerium oxide carbonate hydrate ($Ce_2O(CO_3)_2·H_2O$, reference pattern ICDD card n° 43-0602), as the literature data report that generally lanthanides form basic carbonates in the presence of urea as precipitating agent [9]. Obviously, in the case of binary (or more complex) systems, a mixed cerium–dopant oxide carbonate hydrate is formed [1, 10], even if differential nucleation growths can occur as a result of preexisting concentration gradient or different growth kinetics [3, 9].

The carbonate-based phase thermal decomposition is completed at $T > 320\ °C$, with simultaneous evolution of water and carbon dioxide and consequent formation of the desired fluorite-like cerium-based oxide. This thermal decomposition is paired with a theoretical weight loss of about 23%, according to the following global reaction (for pure cerium oxide):

$$Ce_2O(CO_3)_2 \cdot H_2O + 1/2\,O_2 \rightarrow 2CeO_2 + 2CO_2 + H_2O \tag{4.4}$$

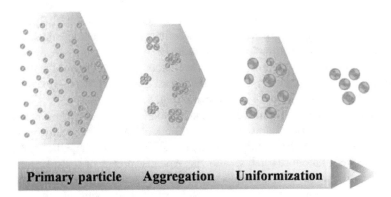

Fig. 4.2 Particles nucleation and aggregation upon UBHP process

According to Eq. 4.4, cerium is oxidized from Ce^{3+} to Ce^{4+} and the desired fluorite-like crystalline structure of cerium oxide is created.

Figure 4.3 shows the thermal evolution of the carbonate-based precursor obtained via UBHP (i.e. ICDD card n° 43-0602) from 200 to 1300 °C.

Fluorite-structured cerium oxide (associated with the ICDD card n° 01-075-0161) began to form at 250 °C [8] (the oxycarbonate phase is co-present along with cubic ceria), and the transformation is completed at 300 °C. Conversely, at 200 °C, the oxycarbonate phase is still the only noticeable phase. Moreover, heating up to 1300 °C did not cause any other structural transformation, although an evident peak sharpening related to irreversible grain growth can still be clearly detected. It is also worth to notice that there is an increasing peak shift toward smaller angles upon treatment temperature increase. That shift is very probably related to material thermal expansion that decreases Bragg angles of the XRD peaks because of the increasing cell parameter. Crystal growth of fluorite-structured doped ceria started from 300 °C and exponentially increased with thermal treatment temperature. It is worth to underline that thermal evolution of $(Ce_2O(CO_3)_2 \cdot H_2O$ is different under different atmospheres, such as a pure nitrogen one. In fact, as reported in [8], the cerium carbonate-based phase begins to decompose at about 430 °C and persists until 500 °C, whereas at $430 < T < 540$ °C an intermediate phase exhibiting an unknown crystal structure is additionally formed. So, under these conditions, the fluorite-like cerium oxide formation is fully completed above 540 °C.

Transmission electron microscopy (TEM) revealed calcined powders morphology. As clearly shown in Fig. 4.4, very small rounded particles (whose size is about few tens of nanometers) together with larger and irregular aggregates of them coexist. Very likely, this morphology is directly derived from the as-synthesized parent powders morphology as it is reported in the literature for many other similar systems [9].

After calcination, the resulting powders exhibit the same morphology of their corresponding parent powders, even if their set of features can be very different according to the adopted synthesis parameters such as solution heating rate, urea concentration, aging time. In all cases, well-defined particles with geometric shapes and very narrow size distribution have always been obtained, very likely thanks to the peculiar advantages of the homogeneous and controlled release of the precipitate ligands in UBHP. Shapes like nanorods, nanospheres, and nanosheets are easily found in the recent literature [3, 8, 11], as shown in Fig. 4.4.

Definitely, UBHP is a promising method to produce nanometric ceria-based powders owning well-defined morphologies and possibly high sinterability, being perfectly suitable for practical applications in IT-SOFCs. In particular, UBHP process is usually applied to produce catalysts allowing a precise precursor control through stable pH conditions, but it can be surely used for different applications such as ceramic electrolytes, thus encouraging further developments and application of UBHP to other systems.

Fig. 4.3 XRD patterns of the as-precipitate GDC10 at various temperatures: **a** 200 °C, **b** 250 °C, **c** 300 °C, **d** 400 °C, **e** 600 °C, **f** 800 °C, and **g** 1300 °C

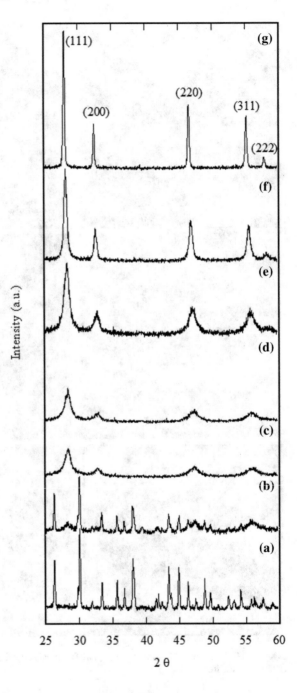

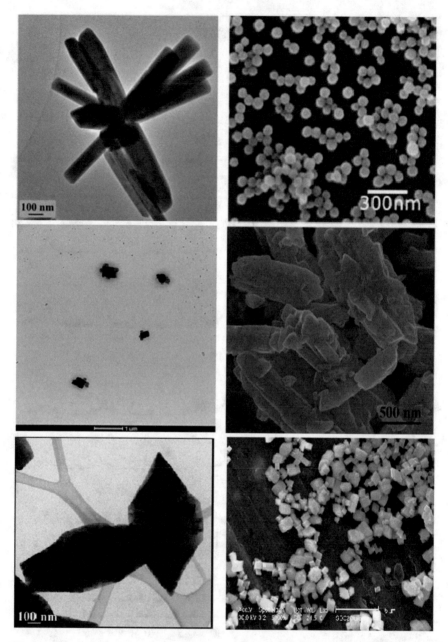

Fig. 4.4 TEM (sx) and SEM (dx) micrographs of variously doped cerium oxide synthesized through UBHP under different conditions

Fig. 4.5 Morphologies of co-precipitated ceria-based powders by using different precipitating agents: ammonia (**a**), TMAH (**b**), ammonium carbonate (**c**), and sodium hydroxide (**d**)

4.3.2 Influence of the Precipitating Agents

In this section, the effects of different precipitating agents (apart from urea, described in Sect. 4.3.1) upon ceria powders properties are analyzed. In particular, the investigated precipitating agents were ammonia (NH_3), tetramethylammonium hydroxide (($CH_3)_4NOH$, i.e. TMAH), ammonium carbonate (($NH_4)_2CO_3$), and sodium hydroxide (NaOH). All those precipitating agents are used in concentrated solution (by using different solvents) at room temperature during the co-precipitation, carried out in either direct or reverse mode. Different precipitating agents have a strong influence upon the formation of the as-precipitate crystalline nature and its morphology.

It is well known that cerium oxide (CeO_2) has a strong tendency to crystallize in the fluorite-like cubic form even at room temperature [12], and this result is also confirmed by the experimental evidences in a wide range of precipitating conditions [13]. In particular, when ammonia, tetramethylammonium hydroxide, or sodium hydroxide are used as precipitating agents for ceria-based systems, the formation of a single-phase CeO_2-based compound is favored [10]. However, differences among the as-precipitates are still clearly evident: Samples synthesized starting from ammonia or TMAH are generally constituted by poorly crystallized fluorite-like ceria [10, 14], whereas samples synthesized by using sodium hydroxide are mostly fully crystallized

[13]. Conversely, when ammonium carbonate is used as precipitating agent, the obtained ceria-based systems are always fully amorphous in case of fast mixing speed [12].

These results are not surprising, anyway. In fact, it is well known that precipitated particles can be either amorphous or crystalline and the factors influencing the final structure are not yet fully clear [15]. For ceria-based systems, crystalline CeO_2 (theoretically without any weight loss) begins to crystallize at 0 °C [16], whereas if $Ce(OH)_4$ (with an associated 17.3% weight loss) is produced after precipitation, it is always amorphous [17]. Therefore, for ceria-based system constituted by both amorphous and crystalline fluorite phase, the larger the weight loss, the more abundant is the residual amorphous content (by assuming that the absorbed water amount is very similar for similarly synthesized powders).

As mentioned before, when ammonium carbonate is used as precipitating agent, a completely amorphous material is produced [18, 19], characterized by large weight losses (in excess to 33%). Although being the phase identification of an amorphous as-precipitate not possible, according to the previous results [20] a reasonable precipitation product in the presence of ammonium carbonate could be $CeOHCO_3 \cdot 2H_2O$, whose total weight loss is 34.0%. Additional and more detailed information about carbonate environment precipitation is given in the following section.

Anyhow, as expected, after calcination step ceria-based as-precipitates are fully crystallized in the fluorite-like structure.

Furthermore, properties shape and powders dimension are also related to different precipitating agents used upon co-precipitation synthesis. Thus, very different particle morphologies and cluster shapes can be obtained by varying precipitating agent under the same synthesis conditions. Additionally, powders synthesized by co-precipitation can form agglomerates, either soft or hard, depending on the adopted precipitating agent and, in turn, on the as-precipitate crystalline nature. Morphologies of differently synthesized powders are reported in Fig. 4.5.

As-precipitated morphologies, reported in Fig. 4.5a, b, d, are similar to each other, being characterized by a broad and irregular distribution of particle size (coexisting compacted particles of several μm along with sub-micrometric particles). Clearly, compaction of such powders is difficult and the resulting green density is very low, thus severely limiting the sintered body final density [15]. Furthermore, a detailed analysis of the larger particles reveals their quite dense structure, very likely corresponding to hard agglomerates. Finally, morphology shown in Fig. 4.5c is similar to the previous ones, although its clusters possess rounded shapes and their structure does not appear dense at all. Very likely, compaction of these powders is easier and the expected green density derived from such powders is significantly higher.

According to the above-presented results, it is possible to claim that for ceria-based ceramics synthesized by co-precipitation, precipitating agents favoring the direct formation of crystalline fluorite ceria lead to hard agglomerates formation after the calcination step. Therefore, to synthesize these materials by co-precipitation route, precipitating agents not favoring the direct crystallization of fluorite-structured doped ceria should be used. In particular, those based on carbonates appear to be a very good production choice.

4.3.3 Carbonate Environment Precipitation

Based on the results described in Sect. 4.3.2 and on interesting literature results [18, 20, 21], ammonium carbonate has been considered as a very good precipitating agent for ceria-based systems, independently from dopant's nature. Actually, nature, size, and morphology of synthesized rare-earth carbonate particles are extremely variable and they depend on manifold factors. The most important ones among them are the carbonate source, the reactants concentration, the synthesis temperature, the reaction time, and the eventual usage of organic ligands. Therefore, by varying all these factors, it is possible to synthesize many different shaped rare-earth carbonates. Hence, the aim of this section is to define the optimal synthesis cycle by analyzing the effects of different parameters upon carbonate environment co-precipitation.

In a typical carbonate environment (i.e. using ammonium carbonate as precipitating agent) precipitation/co-precipitation, a solution containing the desired cations is added to the one containing carbonate anions. The resulting as-precipitate nature and morphology are strongly affected by several synthesis parameters: Firstly, co-precipitation can be carried out either slowly (dripping the cations containing solution at a fixed rate) or quickly (almost instantly); secondly, different R values (being R the molar ratio between ammonium carbonate and rare-earth cations) can be adopted, corresponding either to a slight defect of carbonate ($R \sim 1$ [22]) or to a large excess of carbonate ($R = 10$ [18]); thirdly, different aging temperatures and time lead to strong modifications of the obtained crystalline phases [20, 22].

By considering the influence of different mixing speeds, fast mixing always generates an amorphous precursor, regardless of the other parameters and also in the presence of different co-dopants, very likely because there is not enough time to induce the formation of an ordered structure. According to the results derived mostly from DTA-TG analysis, such amorphous phases correspond to different chemical compositions, even though they are always based on cerium carbonate-/oxycarbonate-related crystalline phases. However, their morphological features are definitely similar to each other, being characterized by favorable spherical shapes and narrow distributions in size. Figure 4.6 highlights the homogeneity and the nanometric nature of various ceria-based particles synthesized by fast carbonate environment co-precipitation. Additionally, Fig. 4.6a shows that single particles tend to agglomerate into bigger and gently rounded clusters, owning the characteristics of soft agglomerates [15] and thus being easily moldable upon forming and sintering steps.

Conversely, slow mixing induces the formation of many different crystalline phases, mainly depending on both R ratio and synthesis/aging temperature. To clarify this tendency, two different cases have been analyzed: the first one, at room temperature with R ratio ranging from 2.5 to 10, and the second one, at higher temperature (i.e. 100 °C) with R ratio ranging from 1 to 5. In the former case, by varying R, two distinct crystalline phases can be obtained, as clearly evident by the XRD patterns shown in Fig. 4.7.

Fig. 4.6 SEM micrographs of amorphous as-precipitates taken at different magnifications, SDC20 (**a**), GDC20 (**b**), CeO$_2$ (**c**), YDC15 (**d**), highlighting both their spherical shape and nanometric particle size

In particular, SDC20R10S XRD pattern exhibits only a single phase, whose peaks can be assigned to ICDD card n° 83-1211, corresponding to an orthorhombic lanthanum–cerium hydrate carbonate [(LaCe)(CO$_3$)$_3$·8H$_2$O].

Anyway, based on the positions and intensities of measured XRD peaks, an orthorhombic samarium–cerium hydrate carbonate is very likely formed and its chemical formula could be (Ce$_{0.8}$Sm$_{0.2}$)$_2$(CO$_3$)$_3$·8H$_2$O. In addition, owning La^{3+}, Ce^{3+} and Sm^{3+} similar values of cationic radius [23], XRD peak positions should be close to those reported in ICDD card n° 83-1211. Conversely, SDC20R2.5S is constituted by two different phases: the same carbonate-based crystalline phase of SDC20R10S and the oxycarbonate-based phase of the ICDD card n° 43-0602 (related to Ce$_2$O(CO$_3$)$_2$·H$_2$O). Even in this case, by considering the present doped ceria system, the adjusted chemical formula of the oxycarbonate-based phase should be (Ce$_{0.8}$Sm$_{0.2}$)$_2$O(CO$_3$)$_2$·H$_2$O. More in detail, the lower ratio between carbonate ions

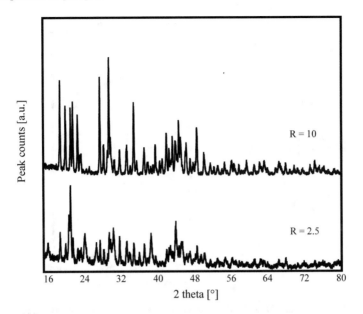

Fig. 4.7 XRD patterns of SDC20 synthesized via carbonate environment co-precipitation by slowly mixing at room temperature and with different R ratios (SDC20R25S and SDC20R10S)

and rare-earth cations dissolved into the co-precipitating solution did not allow a complete formation of samarium-doped cerium carbonate, encouraging on the contrary the formation of the oxycarbonate-based phase characterized by fewer carbonate ions.

SDC20R2.5S and SDC20R10S thermal behavior, reported in Fig. 4.8, very well agrees with the XRD analysis, thus confirming the above-stated hypothesis, as the total weight loss measured for SDC20R10S, i.e. 43%, agrees very well to the theoretical value, i.e. 44.4%, according to the following global reaction:

$$(Ce_{0.8}Sm_{0.2})_2(CO_3)_3 \cdot 8H_2O + 0.4O_2 \rightarrow 2Ce_{0.8}Sm_{0.2}O_{1.9} + 3CO_2 + 8H_2O \quad (4.5)$$

The thermal decomposition of the mixed oxycarbonate-based precursor occurs with the simultaneous formation of the cerium-samarium oxide. Since no other thermal event appears for SDC20R10S, the 20 mol% Sm-doped ceria is formed by the above reaction in the stable crystalline fluorite structure, as it occurs in similar systems too [24].

Therefore, one can assume that in case of slow co-precipitation and $R = 10$, the raw co-precipitate completely crystallizes in $(Ce_{0.8}Sm_{0.2})_2(CO_3)_3 \cdot 8H_2O$, according to the following chemical reaction:

$$1.6Ce_{aq}^{+3} + 0.4Sm_{aq}^{+3} + 3CO_{3\,aq}^{-2} + 8H_2O \rightarrow (Ce_{0.8}Sm_{0.2})_2(CO_3)_3 \cdot 8H_2O_{sol} \downarrow \quad (4.6)$$

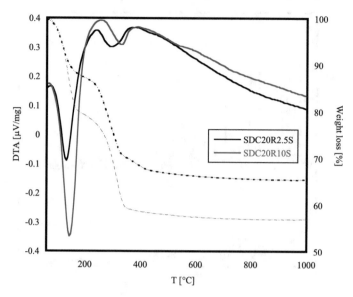

Fig. 4.8 DTA-TG of SDC20 synthesized via carbonate environment co-precipitation by slowly mixing at room temperature and with different R ratios (SDC20R25S and SDC20R10S)

And no residual amorphous phase is present. This reaction is reasonably favored both by a longer available time for precipitation and by a higher concentration of carbonates.

Similarly, the coexistence of two different phases in SDC20R2.5S is also revealed by the corresponding thermal behavior reported in Fig. 4.8. In fact, the first endothermic peak of SDC20R2.5S is similar to the corresponding peak of SDC20R10S, whereas the second endothermic peak is broader and it is located at a lower temperature, i.e. 291 °C instead of 329 °C. Furthermore, two different weight losses appear to be associated with this last peak. This evidence can be ascribed to the evolution of CO_2 from the two different phases. In Fig. 4.8b, the first decomposition of SDC20R2.5S starts at around 200 °C and approximately ends at 320 °C, this being related to the pure carbonate phase. The second decomposition, starting approximately at 320 °C and ending at 500 °C, is related to the oxycarbonate phase decomposition. Moreover, the total weight loss measured for SDC20R2.5S, i.e. about 35%, is a combination of the theoretical weight losses of the two present phases (44.4 and 23.5% for the carbonate and the oxycarbonate phases, respectively). Hence, for $R = 2.5$, the evolution of carbon dioxide, water, and formation of the fluorite phase can be described with the following global reaction:

$$x(Ce_{0.8}Sm_{0.2})_2(CO_3)_3 \cdot 8H_2O + (1 - x)(Ce_{0.8}Sm_{0.2})_2O(CO_3)_2 \cdot H_2O + 0.4O_2 \rightarrow$$
$$2Ce_{0.8}Sm_{0.2}O_{1.9} + (2 + x)CO_2 + (1 + 7x)H_2O \tag{4.7}$$

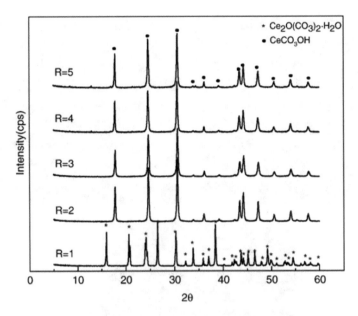

Fig. 4.9 XRD patterns of CeO$_2$ synthesized via carbonate environment co-precipitation by slowly mixing at 100 °C for 12 h with different R ratios (from CEOR1S to CEOR5S)

where x is the mole fraction of carbonate, equal to 0.47 as estimated from the TG-DTA data in the presented case.

In the latter case, again by varying R, two distinct crystalline phases can be obtained, as clearly evident by the XRD patterns shown in Fig. 4.9 [22].

As the solution temperature rises, ammonium carbonate decomposition is favored, thus increasing the local concentration of carbonate anions that react with Ce^{3+} cations to form either the orthorhombic oxycarbonate-based phase of the ICDD card n° 43-0602 (related to Ce$_2$O(CO$_3$)$_2$·H$_2$O) at $R = 1$ (i.e. when the precipitating agent is slightly deficient in the starting solution) or the hexagonal cerium carbonate hydroxide (CeCO$_3$OH, ICDD card n. 62-0031) at $R \geq 2$. Furthermore, as the R ratio increases (i.e. at $R = 5$), the excess of precipitant agent leads to the formation of stable intermediates present as impurities in the related XRD pattern in Fig. 4.9.

The most important effect of different cerium carbonate-/oxycarbonate-based compounds formation consists in their peculiar powder morphologies. In fact, contrary to amorphous as-precipitate, well-crystallized as-precipitates exhibit a wide range of complex morphologies, being their primary particles characterized by geometric and well-defined shapes. As a matter of fact, such features are preserved even after calcination and the consequent formation of stable fluorite-like cerium oxide systems, as for rare-earth carbonate-based materials the as-synthesized powders' morphology and the calcined powders' morphology are very similar [19]. In the present case, each one of the three differently achievable crystalline precur-

(a)   **(b)**

Fig. 4.10 SEM micrographs of the as-precipitates SDC20R10S (**a**), CEOR1S (**b**), CEOR3S (**c**)

sors exhibits its own peculiar morphology, clearly evident by the SEM micrographs reported in Fig. 4.10 for SDC20R10S (pure $(Ce_{0.8}Sm_{0.2})_2(CO_3)_3 \cdot 8H_2O$), CEOR1S (pure $Ce_2O(CO_3)_2 \cdot H_2O$), and CEOR3S (pure $CeCO_3OH$).

SDC20R10S morphology, shown in Fig. 4.10a, is characterized by bi-dimensional hexagonal-like particles, with well-defined geometrical edges, whose length is in the order of several micrometers, width of some micrometers, and thickness of around 1 μm. CEOR1S, shown in Fig. 4.10b, consists in rod-like particles with a very elongated shape, whose length is in the order of several micrometers and whose width is in the order of some hundreds of nanometers. Similar shapes are pretty common for different rare-earth carbonate compounds and have been frequently observed in the literature [12, 19]. Primary particles of both samples shown in Fig. 4.10a, b own unfavorable gave rise to hard agglomerates upon calcination severely impairing their forming and sintering aptitude [15]. Finally, CEOR3S particles, shown in Fig. 4.10c, are spherical in shape and homogeneously distributed in size. Undoubtedly, this set of features is much more advantageous in terms of packing aptitude and, consequently, for the obtaining of well-sintered ceria-based electrolytes, even though this crystalline phase can be synthesized via simple precipitation only under severe synthesis conditions [22].

According to this section and to Sect. 4.3.2, the strong tendency of ceria-based co-precipitates to crystallize very often induces the formation of hardly agglomerated polycrystalline particles with bad morphologies preserved even after calcination. Therefore, the co-precipitation process for ceria-based materials should be as fast as possible, thus impairing or even totally inhibiting cerium compounds crystallization. In fact, any crystalline as-synthesized compound will maintain its morphology even after calcination, thus inducing low green densities onto the formed pellets if such powders have bad packing attitude. If compared to electrolytes derived from well-sintered samples with properly rounded powders morphology without hard aggregates, the final ionic conductivity of electrolytes derived from such bad sintered samples is very low. On the contrary, when the co-precipitated particles are softly agglomerated owning spherical shapes, their sintering behavior is really good, and consequently, the corresponding sintered pellets electrochemical properties are high enough to be efficiently used as electrolytes for IT-SOFCs, even in case of rather complex ceria-based systems [18, 20].

4.4 Hydrothermal Synthesis

Hydrothermal processing is a common synthesis method to obtain nanocrystalline inorganic materials. This synthesis method takes advantage of water solubility of almost all inorganic substances at elevated temperatures and pressures and of the subsequent crystallization of the dissolved material from the fluid [25]. In fact, water at elevated temperatures plays a key role for the precursor transformation: By considering water phase diagram, the area of interest of hydrothermal synthesis relies on the branch between liquid and gaseous water (see Fig. 4.11). In particular, above the water boiling point, all the hydrothermal pressure ranges are available for further considerations.

Generally, hydrothermal synthesis is a simple process to prepare in a single-step homogeneous, easily moldable, and weakly agglomerated powders. In addition, compared to other complex wet chemical synthesis methods, it is rather cheap and easily scalable to produce large quantities of products. Two different operative modes have to be distinguished: autogenous pressure range and high-pressure mode.

In the former case, the process pressure is governed by the liquid-/gas-phase equilibrium. Hence, the operating pressure of the system is given only by temperature, corresponding to the equilibrium line of water phase diagram. Above 374 °C (i.e. 221 bar), gas and liquid phase cannot be distinguished anymore, being the water critical point. Anyway, autogenous pressure at a certain temperature is not always sufficiently high for hydrothermal synthesis necessities, although a great number of inorganic systems (such as ceria-based systems and zirconia-based among them) do not require pressure lying above the critical point.

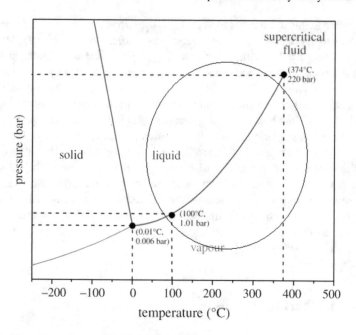

Fig. 4.11 Water phase diagram, highlighting the hydrothermal treatment area of interest

In the latter case, water under a more elevated external pressure rather than the equilibrium water vapor pressure is used. Under these conditions, remarkable results in terms of obtained products can be achieved, but both costs and required experimental setup complexity inevitably increase.

Therefore, water pressure, temperature, and reaction time are the most important physical parameters governing the hydrothermal processing.

The hydrothermal synthesis method is called solvothermal when other solvents rather than water are used during the process. Moreover, it is also possible to add the so-called *mineralizing agents* in order to modify the initial properties of pure hydrothermal water. Polar solvents (i.e. NH_3, CO_2, or different acids or bases to adjust pH) or even nonpolar solvents (i.e. organic compounds, supercritical CO_2) can be used in order to extend the operative range of this synthesis method for the dissolution/recrystallization process underwent by the precursor. Although presenting several advantages, these solvents have also specific drawbacks such as toxicity or corrosion problems for the autoclave materials. Finally yet importantly, most of the original process simplicity is lost by using them.

4.4.1   Carbonate Environment Hydrothermal Synthesis

According to the renewed interest toward rare-earth carbonates as precursors and sacrificial templates for obtaining functional oxides [19], ceria-based carbonates/oxycarbonates have been deeply investigated for their easily tuneable morphology, size, and shape. In particular, one of the easiest and common ways to synthesize such ceria-based carbonate compounds is the hydrothermal/solvothermal route, due to its relatively low costs, simplicity, and industrial potential scale-up. The reaction time and duration, as well as source and concentration of both rare-earth cations and carbonate anions, solvent used, and pH of the reaction solution play a crucial role in controlling the as-obtained particles characteristics.

Generally, for ceria-based systems, hydrothermal treatments are carried out over a range of 100–240 °C and for 1–72 h [19], whereas the most common carbonate sources are: urea [26–28], ammonium carbonate [12, 19], sodium carbonate [19], and sodium hydrogen carbonate [19].

Rare-earth carbonate compounds and, in particular, the ceria-based ones are able to crystallize in many different structures. Among them, the most frequently synthesized structures at nano-/microscale are: rare-earth (RE) hydroxycarbonates (i.e. $REOHCO_3$), oxycarbonates hydrates (i.e. $RE_2O(CO_3)_2 \cdot H_2O$), dioxycarbonates ($RE_2O_2CO_3$), and carbonate octahydrate (i.e. $RE_2(CO_3)_3 \cdot 8H_2O$). Different crystalline structures can be obtained by varying manifolds parameters of the hydrothermal synthesis, even if the most affecting parameters appear to be both carbonate source/concentration and the reaction time.

Keeping in mind that ceria-based carbonate compounds have to be used (according to the focus of this book) as precursors for dense solid electrolytes and that rare-earth carbonate compounds always maintain their morphologies upon decomposition [19, 20, 28], the as-synthesized powders have to be characterized by size, shape, and features maximizing their molding and sintering aptitudes [15]. More in detail, such precursor powders have to be nanometric in size (with homogeneous size distribution), regular in shape (possibly spherical-like), and lowly agglomerated. Generally, hexagonal cerium hydroxycarbonate (i.e. $CeOHCO_3$, ICDD card n. 62-0031) exhibits this whole set of features [12, 28]. Accordingly, the following discussion will focus on obtaining such peculiar crystalline phase and, in particular, on the use of either urea or ammonium carbonate as mineralizing agents for the hydrothermal treatments, as they are the ones favouring its formation.

By considering urea as carbonate source, the hexagonal cerium hydroxycarbonate is obtained as a result of phase transformations of the firstly formed crystalline phases, strongly depending on the chosen reaction temperature. A possible formation mechanism of the oxycarbonate-based phase in the presence of urea as mineralizing agent has been proposed by Cui et al. [28], involving the kinetics of urea hydrolysis producing both ammonium and cyanate ions.

Figure 4.12 shows the XRD patterns of two different Sm-doped ceria-based systems hydrothermally treated at 120 °C and at different reaction times, i.e. 8 and 48 h (labeled as SDC20U8 and SDC20U48, respectively).

SDC20 powders are constituted by two different crystalline phases, depending on the treatment duration. The former phase, produced during the first few hours of hydrothermal treatment, is an orthorhombic structure of cerium hydroxycarbonate. In particular, SDC20U8 is constituted by only this peculiar phase, as all its peaks are attributable to ICDD card n. 41-0013 (see Fig. 4.12), although a slight shift suggesting the presence of Sm^{3+} partially substituting Ce^{4+} ions. With the increasing of treatment time, SDC20 samples undergo phase transformation with consequent formation of a second crystalline phase, according to dissolution/recrystallization processes typical of the hydrothermal synthesis [29]. As evident from Fig. 4.12, SDC20U48 exhibits peaks attributable to the hexagonal cerium carbonate hydroxide (i.e. $CeCO_3OH$, ICDD card n. 62-0031) along with the oxycarbonate-based phase peaks belonging to ICDD card n. 41-0013. Samples treated at reaction times ranging from 8 to 48 h show an intermediate behavior, as their XRD patterns follow a crescent trend with increasing times from the oxycarbonate-based phase to the hexagonal carbonate-based phase peaks intensity.

Figure 4.13 shows four exemplary SEM micrographs of SDC20 samples synthesized by using urea as mineralizing agent at 120 °C, highlighting the morphology evolution of the as-synthesized powders upon treatment time, i.e. 8, 16, 24, 48 h.

As expected from the previous results [28], the oxycarbonate-based phase of SDC20U8 (Fig. 4.13a) is characterized by a peculiar flower-like morphology that is unfavorable for a good powder sintering attitude, as it well known that even after calcination powders derived from rare-earth carbonates preserve their parent mor-

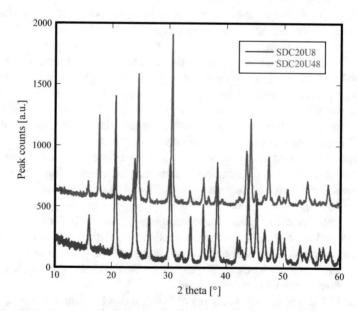

Fig. 4.12 XRD patterns of SDC20U8 and SDC20U48, hydrothermally synthesized by using urea as mineralizing agent at 120 °C and at different reaction times

phologies [19]. Still, few spherical particles are visible in Fig. 4.13a, very likely related to small percentages (undetected by XRD measurements) of the hexagonal oxycarbonate-based phase (i.e. ICDD card n. 62-0031) that began to form after the first hours of the hydrothermal treatment.

Following the evolution of powders crystalline structure and morphology, it is possible to note that with the increase of treatment time, flower-like structures are progressively lost, as the elongated primary particles coexist with spherical particles within the same agglomerates. After 48 h of treatment (SDC20U48 in Fig. 4.13d), rounded particles related to the hexagonal oxycarbonate-based phase (i.e. ICDD card n. 62-0031) are the majority, even if the residual phase evolution is far from complete.

As it has been reported that the hexagonal $CeCO_3OH$ structure is more stable than the orthorhombic structure at higher temperatures [30], different results can be obtained by simply increasing treatment duration under the same synthesis conditions.

Figure 4.14 shows the XRD patterns of two different Sm-doped ceria-based systems hydrothermally treated at different temperatures (i.e. 120, 150, and 160 °C) for 48 h, highlighting the temperature-dependant transition from the orthorhombic structure of doped $CeCO_3OH$ to the hexagonal one.

Fig. 4.13 SEM micrographs of SDC20 hydrothermally synthesized by using urea as mineralizing agent at 120 °C and at different reaction times, SDC20U8 (**a**), SDC20U16 (**b**), SDC20U24 (**c**), and SDC20U48 (**d**)

All the diffraction peaks shown in Fig. 4.14 can be assigned to the orthorhombic CeCO₃OH structure (ICDD card n. 41-0013) and to the hexagonal $CeCO_3OH$ structure (ICDD card n. 62-0031). As also evident from Fig. 4.13, SDC20U48 (pattern A in Fig. 4.14) synthesized at 120 °C is constituted by a combination of both crystalline phases, even if with the reaction temperature increases to 150 ₀C, the ratio between the orthorhombic and the hexagonal $CeCO_3OH$ decreases (pattern B in Fig. 4.14). Finally, at 160 °C as treatment temperature, the SDC20U48 sample is almost completely composed by the hexagonal cerium oxycarbonate (pattern C in Fig. 4.14).

The effect of the increasing treatment temperature upon the as-synthesized particles morphology is negligible, whereas their size and degree of agglomeration are more evident [28].

The hexagonal cerium carbonate hydroxide ($CeCO_3OH$, ICDD card n. 62-0031) can also be easily produced via hydrothermal treatment by using ammonium carbonate as mineralizing agent/carbonate source. Similar to urea environment, powder precursors undergo different transformations upon the hydrothermal treatment, depending on both duration and adopted temperature. Additionally, by using ammonium carbonate as carbonate source, R ratio between carbonate anions and rare-earth cations plays a crucial role in the formation of the hexagonal $CeCO_3OH$. For the sake of comparison, the following discussion refers to 120 °C as treatment temperature with times ranging from 8 to 48 h.

Figure 4.15 shows the XRD patterns of both precursor and hydrothermally treated samples at three different times (all related to $R = 2.5$).

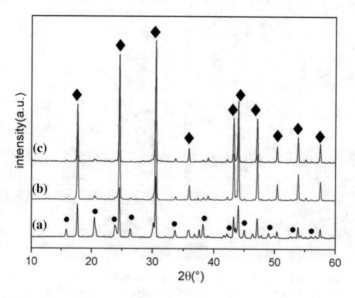

Fig. 4.14 XRD of SDC20U48, hydrothermally synthesized by using urea as mineralizing agent at 120 °C (**a**), 150 °C (**b**), 160 °C (**c**); Diagonal symbol-hexagonal, Filled circle-orthorhombic

Fig. 4.15 XRD patterns of
SDC20 at $R = 2.5$ (**a**) and
after hydrothermal treatment
of 8 h (**b**), 16 h (**c**), and 24 h
(**d**)

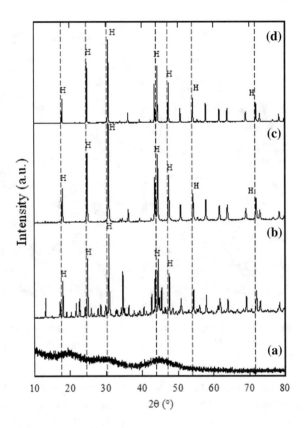

As expected from evidences reported in Sects. 4.3.2 and 4.3.3, the raw precursor is fully amorphous at room temperature, and in particular, it should correspond to a hydrated samarium-doped cerium hydroxide whereof the chemical formula is $(Ce_{0.8}Sm_{0.2})OHCO_3 \cdot 2H_2O$. Conversely, by considering samples obtained after 16 h of hydrothermal treatment at 120 °C, the amorphous precursor completely crystallizes (Fig. 4.15c), as all its peaks can be assigned to the hexagonal $CeCO_3OH$ (labeled with "H" in Fig. 4.15). By carefully analyzing the XRD patterns, a slight shift in the main XRD peak positions with respect to the ones reported in ICDD card n. 62-0031 (represented by the dashed lines in Fig. 4.15) is present, thus suggesting that Ce^{4+} cation is partially substituted with Sm^{3+} cation as the measured XRD peaks are all slightly shifted forward [23].

XRD patterns belonging to samples treated for 8 and 24 h are shown in Fig. 4.15b, d, respectively. In the latter case, SDC20 powders consist in the same above-mentioned hexagonal hydroxide carbonate phase but with larger crystals, as revealed by narrower and sharper XRD peaks. Conversely, in the former case a much more complex diffraction pattern appears, characterized by the presence of hexagonal hydroxide carbonate along with at least another crystalline phase.

Fig. 4.16 TEM micrographs of SDC20 hydrothermally synthesized at 120 °C for 8 h (**a, b**), 16 h (**c, d**), and 48 h (**e, f**)

The hexagonal phase deriving from the hydrothermal treatment with ammonium carbonate confirms having a homogeneously spherical morphology, being nanometric in size and with a rather narrow size distribution as it clearly appears in the corresponding TEM micrographs (taken at different magnifications) reported in Fig. 4.16. Conversely, the rod-like morphology reported in Fig. 4.16 too is associated with the unidentified mixed cerium–samarium carbonate. Indeed, it is evident that reaction time affects not only the obtained crystalline phases but also their morphologies.

Figure 4.16a, b shows TEM micrographs of sample obtained after 8 h of hydrothermal treatment. According to the corresponding XRD pattern (Fig. 4.15b), the presence of two kinds of particles (and related crystalline phases) is confirmed.

The first one is characterized by acicular particles with a very elongated shape, whose length is in the order of several micrometers and whose width is in the order of some hundreds of nanometers. Zhang et al. [31] identified for a similar system particle with the same morphology as orthorhombic $Y_2(CO_3)_3 \cdot 2.5H_2O$ (tengerite-(Y), ICDD card n. 24-1419). Vallina et al. [32] found that, by hydrothermal treatment of $Nd(NO_3)_3$ in the presence of Na_2CO_3, particles with similar morphology were formed and these crystals were identified as tengerite (Nd) with chemical formula $Nd_2(CO_3)_3 \cdot 2.5H_2O$. Unfortunately, in both ICDD and ICSD databases an analogous orthorhombic cerium carbonate does not exist. Actually, an orthorhombic hydrated cerium-based carbonate could be the best candidate for this particular crystalline phase.

Moreover, by considering the micrographs in Fig. 4.16c, d, virtually all SDC20R2.5 particles own a spherical shape. Obviously, the ideal particles shape to produce a dense ceramic electrolyte is the spherical one because of their better packing homogeneity, whereas powders owning the morphology of Fig. 4.16a, b cannot very likely achieve an adequate final densification even at high sintering temperatures, and so their final properties are severely impaired. Finally, further increase in treatment times (i.e. 48 h and more) leads to many fragmentary nanorods-shaped particles, as evident in Fig. 4.16e, f. This result is probably related to new growth units/renucleation, leading to the formation of new recurring morphologies.

Finally, to highlight the effect of the R ratio in ammonium carbonate environment, Fig. 4.17a, b shows analogous SDC20 samples synthesized by hydrothermal treatments at 120 °C in the presence of a large excess of carbonate, i.e. $R = 10$. It is evident that carbonate ions/cations ratio is crucial to establish the nature of the final products. In particular, due to the large excess of carbonate ions, the cerium carbonate-based compound (possibly a tengerite-type structure) is obviously favoured with respect to the hexagonal cerium hydroxide carbonate ($CeCO_3OH$). Not surprisingly, only the acicular particles are present at these conditions. As expected, such morphology is preserved even after the calcination step that is needed to induce the formation of fluorite-structured doped cerium oxide, as evident in Fig. 4.17c, d.

As predictable, powders derived from $CeCO_3OH$ precursors exhibit a very good sintering behavior, achieving very high values of relative density [12, 33], even at significantly lower temperatures as usually adopted for ceria-based systems. In fact, all those specimens generally present a highly homogeneous microstructure, characterized by narrow grain size distributions in the nano/micrometric range. On the contrary, when crystalline precursors own bad-quality morphologies such as the ones showed in Figs. 4.13a, 4.16a, b or 4.17c, d, the deriving powders are hard to pack and, in turn, their sintering behavior is impaired.

Rather than with urea or ammonium carbonate, $CeCO_3OH$ can also be synthesized via hydrothermal treatments at more complex conditions; for example, Gao et al. [34] obtained hexagonal single-crystalline cerium carbonate hydroxide precursors at 180 °C using $CeCl_3 \cdot 7H_2O$ as cerium source and triethylenetetramine as carbonate source; Fan et al. [35] synthesized crystalline $CeCO_3OH$ at 120 °C and carbamide as mineralizing agent. However, the simplest way to produce such crystalline structure with nearly ideal packing morphology consists in opportunely tuning treatment temperature and duration (R ratio in case of ammonium carbonate too) by using either urea or ammonium carbonate (being both very cheap and available reagents) as carbonate source/mineralizing agent.

In conclusion, this section aimed to explain the phenomena involved in the carbonate environment of hydrothermal synthesis and, in particular, in obtaining crystalline cerium-based carbonate hydroxide, with hexagonal structure, corresponding to the best carbonate precursor for doped ceria with optimal sintering aptitude, regardless of the complexity of the system. Definitely, these guidelines could be a real advancement in the production of ceria-based electrolytes via scalable hydrothermal synthesis processes.

Fig. 4.17 TEM micrographs of as-synthesized (**a**, **b**) and calcined (**c**, **d**) SDC20R10, hydrothermally synthesized at 120 °C for 8 h

4.5 Sol-gel Synthesis

Sol-gel processing is a promising method for the preparation of nanostructured materials. The reaction product of the sol-gel synthesis is either colloidal powders or thin films. The main advantage of this synthesis method is the ability to control the final microstructure of the obtained powders by controlling chemical reaction parameters. In fact, it has been proved that modification of the reaction conditions could substantially affect the final solgel product features [36, 37]. In fact, this method is widely used for the synthesis of inorganic and organic/inorganic hybrid materials, being capable of producing either nanoparticles, nanorods, thin films, or monoliths. Also for ceria-based systems, sol-gel synthesis has been successfully applied to realize easily sinterable nanosized powders [38].

However, sol-gel synthesis has also its own disadvantages, such as the high cost of precursors (usually metal alkoxides), the low rate of solvent removal from the obtained gel, and the always needed subsequent thermal treatment. Moreover, gel formation can be a slow process, thus making sol-gel method a time-consuming fabrication technique compared to other commonly used methods (Fig. 4.18).

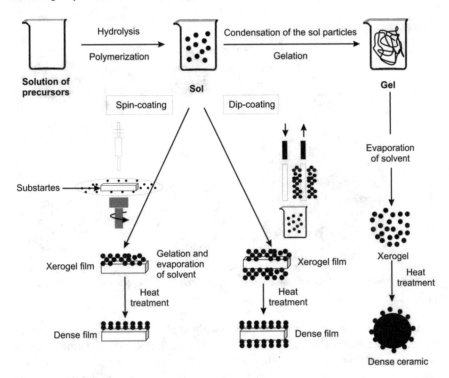

Fig. 4.18 Examples of differently modified sol-gel synthesis

In some cases, modified sol-gel syntheses could lead to different classes of materials such as aerogels and ambigels [39] constituted by continuous networks of pores and solids that can be highly beneficial for many particular applications. In particular, a well-connected pore network favors a quick diffusional transport of gaseous reactants and/or products [40], and the combination of high surface area, porosity, and bonded solid networks promotes very high electronic and electrical conduction. This latter set of feature could be particularly advantageous for ceria-based electrodes rather than electrolytes, and significant efforts to investigate such nanoceria architectures for functional SOFC and SOEC applications have still to be done [39].

4.6 Microwave-Assisted Synthesis

The interaction of dielectric materials, either liquids or solids, with microwaves is the so-called dielectric heating. More in detail, electric dipoles present in dielectric materials react to an externally applied electric field. In liquids, constant reorientation leads to molecular friction with consequent heat generation. However, microwave irradiation effects include both thermal and nonthermal effects [41]. Microwave

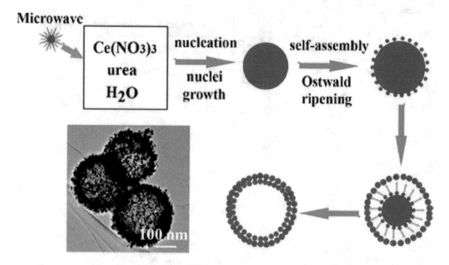

Fig. 4.19 Synthetic route for spherical ceria particles through microwave-assisted synthesis

irradiation as a heating method has been widely applied in chemistry to synthesize nanostructured materials. The microwave ovens used in microwave chemistry range from simple multimode ovens to large-scale batch as well as continuous multimode ovens. Compared to conventional oven heating (i.e. with slow heating rate and heat transfer), microwave heating of liquids constitutes an alternative heating approach with specific advantages.

In particular, microwave-assisted heating provides a more rapid and simultaneous environment for nanoparticles formation thanks to the fast and homogeneous heating effects of microwave irradiation. Hence, microwave-assisted synthesis presents several advantages such as short reaction time, high energy efficiency, and the ability to induce the formation of nanometric particles with narrow size distribution and high chemical purity.

Microwave-assisted synthesis of inorganic compounds has been studied since the 1995, and its interest keeps growing in its application to materials science [42]. In particular, microwave-assisted synthesis is often being used to synthesize functional nanomaterials and rare-earth carbonate precursors [43]. Apart from the above-mentioned generic advantages, for ceria-based systems in carbonate environments, the microwave-assisted synthesis affects particles morphology and their growth. In fact, by comparing a microwave-assisted hydrothermal treatment with a conventional one involving cerium nitrate and urea under the same conditions (i.e. temperature and treatment duration), significant differences in terms of as-obtained powders morphology can be observed [19], even if the powders crystalline habitus was exactly the same (i.e. $Ce_2O(CO_3)_3$). More in detail, microwave heating promoted the formation of spherical particles whereas conventional heating favored the formation of irregular rod-like particles whose length was affected by different urea concentrations. Definitely, high-quality ceria-based products could be obtained through microwave-assisted synthesis by simply varying manifold starting parameters. Figure 4.19 shows a possible mechanism for the formation of hollow ceria nanospheres according to [21] and starting from similar synthesis conditions of [19, 43].

References

1. G. Accardo, L. Spiridigliozzi, R. Cioffi, C. Ferone, E.D. Bartolomeo, S.P. Yoon, G. Dell'Agli, Gadolinium-doped ceria nanopowders synthesized by urea-based homogeneous co-precipitation (UBHP). Mater. Chem. Phys. **187**, 149–155 (2017)
2. W.H.R. Shaw, J.J. Bordeaux, The decomposition of urea in aqueous media. J. Am. Chem. Soc. **77**(18), 4729–4733 (1955)
3. H. Qin, X. Tan, W. Huang, J. Jiang, H. Jiang, Application of urea precipitation method in preparation of advanced ceramic powders. Ceram. Int. **41**(9), 11598–11604 (2015)
4. X. Xu, X. Sun, H. Liu, J.G. Li, X. Li, D. Huo, S. Liu, Synthesis of monodispersed spherical Yttrium Aluminum Garnet (YAG) powders by a homogeneous precipitation method. J. Am. Ceram. Soc. **95**(12) (2012)
5. Z. Mohammadi, A.S.M. Mesgar, F. Rasouli-Disfani, Preparation and characterization of single phase, biphasic and triphasic calcium phosphate whisker-like fibers by homogeneous precipitation using urea. Ceram. Int. **42**(6), 6955–6961 (2016)
6. Y.S. Wu, Y.H. Lee, H.C. Chang, Preparation and characteristics of nanosized carbonated apatite by urea addition with coprecipitation method. Mater. Sci. Eng., C **29**(1), 237–241 (2009)
7. J. Subrt, V. Stengl, S. Bakardijeva, L. Szatmary, Synthesis of spherical metal oxide particles using homogeneous precipitation of aqueous solutions of metal sulfates with urea. Powder Technol. **169**(1), 33–40 (2006)
8. A.M. D'Angelo, N.A.S. Webster, A.L. Chaffee, Characterisation of the phase-transformation behaviour of $Ce_2O(CO_3)_2*H_2O$ clusters synthesised from $Ce(NO_3)_3*6H_2O$ and urea. Powder Diffr. **29**(S1), S84–S88 (2014)
9. J. Li, X. Li, X. Sun, T. Ikegami, T. Ishigaki, Uniform colloidal spheres for $(Y_{1−x}Gd_x)_2O_3$ ($x = 0–1$): formation mechanism, compositional impacts, and physicochemical properties of the oxides. Chem. Mater. **20**(6), 2274–2281 (2008)
10. L. Spiridigliozzi, G. Dell'Agli, M. Biesuz, V.M. Sglavo, M. Pansini, Effect of the precipitating agent on the synthesis and sintering behavior of 20 mol% Sm-doped ceria. Adv. Mater. Sci. Eng. **2016**, 1–8 (2016)
11. A.M. D'Angelo, A.L. Chaffee, Correlations between oxygen uptake and vacancy concentration in Pr-doped CeO_2. ACS Omega 2544–2551 (2017)
12. G. Dell'Agli, L. Spiridigliozzi, A. Marocco, G. Accardo, D. Frattini, Y. Kwon, S.P. Yoon, Morphological and crystalline evolution of Sm-(20 mol%)–doped ceria nanopowders prepared by a combined co-precipitation/hydrothermal synthesis for solid oxide fuel cell applications. Ceram. Int. **43**, 12799–12808 (2017)
13. M.J. Godinho, R.F. Goncalves, L.P.S. Santos, J.A. Varela, E. Longo, E.R. Leite, Room temperature co-precipitation of nanocrystalline CeO_2 and $Ce_{0.8}Gd_{0.2}O_{1.9-d}$ powder. Mater. Lett. **61**, 1904–1907 (2007)
14. M. Mazaheri, S.A.H. Tabrizi, M. Aminzare, S.K. Sadrnezhaad, Synthesis of CeO_2 nanocrystalline powder by precipitation method. Ceram. Mater. **62**(4), 529–532 (2010)
15. M.N. Rahaman, *Ceramic Processing and Sintering* (CRC Press, 2003)
16. C.J. Shih, Y.J. Chen, M.H. Hon, Synthesis and crystal kinetics of cerium oxide nanocrystallites prepared by co-precipitation process. Mater. Chem. Phys. **121**(1–2), 99–102 (2010)
17. Y.X. Li, X.Z. Zhou, Y. Wang, X.Z. You, Preparation of nano-sized CeO_2 by mechanochemical reaction of cerium carbonate with sodium hydroxide. Mater. Lett. **58**(1–2), 245–249 (2004)

18. L. Spiridigliozzi, G. Dell'Agli, A. Marocco, M. Pansini, G. Accardo, S.P. Yoon, H.C. Ham, D. Frattini, Engineered co-precipitation chemistry with ammonium carbonate for scalable synthesis and sintering of improved $Sm_{0.2}Ce_{0.8}O_{1.90}$ and $Gd_{0.16}Pr_{0.04}Ce_{0.8}O_{1.90}$ electrolytes for IT-SOFCs. J. Ind. Eng. Chem. **59**, 17–27 (2018)

19. A.M. Kaczmarek, K.V. Hecke, R.V. Deun, Nano- and micro-sized rare-earth carbonates and their use as precursors and sacrificial templates for the synthesis of new innovative materials. Chem. Soc. Rev. **44**(8), 2023–2576 (2015)

20. J.G. Li, T. Ikegami, Y. Wang, T. Mort, Reactive ceria nanopowders via carbonate precipitation. J. Am. Ceram. Soc. **85**(9), 2376–2378 (2002)

21. D. Zhang, X. Du, L. Shi, R. Gao, Shape-controlled synthesis and catalytic application of ceria nanomaterials. Dalton Trans. **41**, 14455–14475 (2012)

22. M.Y. Cho, K.C. Roh, S.M. Park, H.J. Choi, J.W. Lee, Control of particle size and shape of precursors for ceria using ammonium carbonate as a precipitant. Mater. Lett. **64**(3), 323–326 (2010)

23. R.D. Shannon, Revised effective ionic radii and systematic studies of interatomic distances in halides and chalcogenides. Acta Crystallogr. A 751–767 (1976)

24. D.W. Joh, M.K. Rath, J.W. Park, J.H. Park, K.H. Cho, S. Lee, K.J. Yoon, J.H. Lee, K.T. Lee, Sintering behavior and electrochemical performances of nano-sized gadolinium-doped ceria via ammonium carbonate assisted co-precipitation for solid oxide fuel cells. J. Alloy. Compd. **682**, 188–195 (2016)

25. O. Schaf, H. Ghobarkar, P. Knauth, Hydrothermal synthesis of nanomaterials, in *Nanostructured Materials* (Kluwer Academic Publishers, 2004), pp. 23–42

26. Z.H. Han, N. Guo, K.B. Tang, S.H. Yu, H.Q. Zhao, Y.T. Qian, Hydrothermal crystal growth and characterization of cerium hydroxycarbonates. J. Cryst. Growth **219**(3), 315–318 (2000)

27. F. Hrizi, H. Dhaouadi, F. Dhaouadi, Cerium carbonate hydroxide and ceria micro/nanostructure: synthesis, characterization and electrochemical properties of $CeCO_3OH$. Ceram. Int. **40**, 25–30 (2014)

28. M.Y. Cui, J.X. He, N.P. Lu, Y.Y. Zheng, W.J. Dong, W.H. Tang, B.Y. Chen, C.R. Li, Morphology and size control of cerium carbonate hydroxide and ceria micro/nanostructures by hydrothermal technology. Mater. Chem. Phys. **121**, 314–319 (2010)

29. F. Meng, C. Zhang, Z. Fan, J. Gong, A. Li, Hydrothermal synthesis of hexagonal CeO_2 nanosheets and their room temperature ferromagnetism. J. Alloy. Compd. **647**, 1013–1021 (2015)

30. M. Hirano, E. Kato, Hydrothermal synthesis of nanocrystalline cerium(IV) oxide powders. J. Am. Ceram. Soc. **82**(3) (2004)

31. Y. Zhang, M. Gao, K. Han, Z. Fang, X. Yin, Z. Xu, Synthesis, characterization and formation mechanism of dumbbell-like $YOHCO_3$ and rod-like $Y_2(CO_3)_3$ $2.5H_2O$. J. Alloy. Compd. **474**, 598–604 (2009)

32. B. Vallina, J.D. Rodriguez-Blanco, A.P. Brown, J.A. Blanco, L.G. Benning, The role of amorphous precursors in the crystallization of La and Nd carbonates. Nanoscale **7**, 12166–12179 (2015)

33. M.Y. Cheng, D.H. Hwang, H.S. Sheu, B.J. Hwang, Formation of $Ce_{0.8}Sm_{0.2}O_{1.9}$ nanoparticles by urea-based low-temperature hydrothermal process. J. Power Sources **175**, 137–144 (2008)

34. K. Gao, Y.Y. Zhu, D.Q. Tong, L. Tian, Z.H. Wang, X.Z. Wang, Hydrothermal synthesis of single-crystal $CeCO_3OH$ and their thermal conversion to CeO_2. Chin. Chem. Lett. **25**, 383–386 (2014)

35. T. Fan, L. Zhang, H. Jiu, Y. Sun, G. Liu, Y. Sun, Q. Su, Template-free hydrothermal synthesis and characterisation of single crystalline $Ce(OH)CO_3$ and CeO_2 with spindle-like structures. IET Micro Nano Lett. **5**(4), 230–233 (2010)

36. A. Dupont, C. Parent, B.L. Garrec, J.M. Heintz, Size and morphology control of Y_2O_3 nanopowders via a sol-gel route. J. Solid State Chem. **171**(1–2), 152–160 (2003)

37. A. Sutka, G. Mezinskis, Sol-gel auto-combustion synthesis of sinel-type ferrite nanomaterials. Front. Mater. Sci. **6**(2), 128–141 (2012)

38. W. Huang, P. Shuk, M. Greenblatt, Properties of sol-gel prepared $Ce_{1-x}Sm_xO_{2-x/2}$ solid electrolyte. Solid State Ionics **100**(1–2), 23–27 (1997)
39. C. Laberty-Robert, J.W. Long, E.M. Lucas, K.A. Pettigrew, R.M. Stroud, M.S. Doescher, D.R. Rolison, Sol-gel-derived ceria nanoarchitectures: synthesis, characterization, and electrical properties. Chem. Mater. **18**, 50–58 (2006)
40. J.M. Wallace, J.K. Rice, J.J. Pietron, R.M. Stroud, J.W. Long, D.R. Rolison, Silica nanoarchitectures incorporating self-organized protein superstructures with gas-phase bioactivity. Nano Lett. **3**(10), 1463–1467 (2003)
41. A.G. Saskia, Microwave chemistry. Chem. Soc. Rev. **3** (1997)
42. Y. Chiang, C. Shih, A. Sie, M. Li, C. Peng, P. Shen, Y. Wang, T. Guo, P. Chen, Highly stable perovskite solar cells with all-inorganic selective contacts from microwave-synthesized oxide nanoparticles. J. Mater. Chem. A (In press) (2017)
43. Y. Ikuma, H. Oosawa, E. Shimada, M. Kamiya, Effect of microwave radiation on the formation of $Ce_2O(CO_3)_2*H_2O$ in aqueous solution. Solid State Ionics **151**(1–4), 347–352 (2002)

Chapter 5
Doped Ceria Electrolytes: Alternative Sintering Methods

5.1 Flash Sintering (FS)

Flash sintering consists in sintering a material by simultaneously exposing it to an electric field and to heat [1]. In the FS process, when a critical combination of electric field and outset temperature is reached, a power surge occurs (the so-called *flash event*), resulting in a nearly instantaneous full densification of the exposed material [2]. Flash event occurs after an incubation time depending upon field strength and flash temperature.

Electrical conductivity is a key parameter in the FS process, and it can be used to differentiate materials behavior toward FS. According to the recent literature [3], FS has been applied to a wide range of metallic conductors, ionic conductors, semiconductors, and room temperature insulators. Even if conductivity has been deeply studied during the last century, at present it is not fully understood how field strength (combined with sample heating) affects the conductivity of materials during the FS process. For applied voltages up to tens of kV, oxide insulators exhibit a nonlinear Ohmic increase in conductivity as a function of the applied electric field. Conductivity-related mechanisms might not be strictly applicable to FS, being usually studied only at room temperature. Higher temperatures, typically around the onset temperatures of FS, are sufficient to promote complex ionic conditions and other effects such as electrochemical reduction [3]. Hence, the involved complexity is rather high, the electrical conductivity of oxides spanning over more than 22 orders of magnitude between insulating, semiconducting, and metallic behavior.

The conductivity temperature dependence plays, very likely, a key role during the FS process, affecting the electrical resistivity of materials. In particular, when temperature increases, the electric conductivity of ionic conductors (i.e. YSZ and doped ceria) and insulators (i.e. $BaTiO_3$) rapidly increases, while the conductivity of semiconductors (i.e. SiC and B_4C) increases at lower rates. Conversely, conductivity of metallic materials (i.e. W and Cu) slightly decreases with the increase of operating temperature.

L. Spiridigliozzi, *Doped-Ceria Electrolytes*, SpringerBriefs in Applied Sciences and Technology, https://doi.org/10.1007/978-3-319-99395-9_5

Hence, the FS mechanisms driving the rapid densification are still an open topic. Without electric currents (i.e. conventional sintering), well-established sintering theories have been developed. Upon conventional sintering, six mechanisms have been identified: lattice diffusion, grain boundary diffusion, viscous flow, surface diffusion, and gas-phase transport. On the contrary, sintering theories involving the electric field contribution in the governing equations are still under development. Anyway, these material transport mechanisms may depend on conductivity modes (i.e. ionic, electronic, or mixed), polarity-induced effects, current densities, and applied voltages. Either several theories based on experimental evidence or hypothesized mechanisms have been proposed to explain the ultra-rapid densification. They can be classified into: (I) extremely rapid macroscale Joule heating with consequent heat localization at lattice scale, (II) Frenkel pairs nucleation with consequent formation of vacancies, (III) complex electrochemical reduction, even though the first one is the most accepted mechanism, being applicable to materials owning any kind of conductivity mode.

5.1.1   A Comparison Between Conventional and Field-Assisted Sintering of Gadolinium-Doped Ceria

The present section aims to deeply analyze the effect of flash sintering (FS) upon ceria-based systems through the optimization of powder composition and operative conditions. In particular, FS mechanism can be analyzed by considering the electronic conductivity of materials and their effect on flash sintering. Firstly, Gd-doped ceria and Sm-doped ceria were selected as test materials as they have high ionic conductivity with appropriate electronic conductivity [4]. Additionally, the electronic conductivity contribution of rare-earth-doped ceria can be altered by changing the doping element and the doping amount, to unravel the relationship between flash sintering onset and the related starting powder conductivity [5]. Materials considered for flash sintering in this section are pure ceria (named C in the following), GDC5, GDC10, GDC15, GDC20, and SDC20. Furthermore, several GCD10 samples were also prepared by adding 1 mol% Li_2O or CoO as sintering aids (LiGDC10 and CoGDC10 in the following) [6].

Flash sintering was performed in a specifically modified dilatometer using a constant heating rate of 20 °C/min. Two platinum plates acted as electrodes, pressed against the sample with a constantly applied load. To improve the electrical contact between platinum electrodes and specimens, silver paste was applied to the green pellet upper and lower surfaces. Various electric fields were applied by using a DC power supply, switched on when the sample temperature reached 150 °C. A multimeter monitored the electrical parameters (voltage and current). Once the maximum current limit was reached, the system operated under current control for few minutes before the final setup shuts down.

Figure 5.1 shows a schematic representation of the adopted experimental setup for flash sintering.

Two different powder precursors will be considered in the following discussion, i.e. commercial-grade-doped ceria powders [7] and as-synthesized doped ceria powders via co-precipitation method [5, 6]. Different powders reactivity (related to different particle size and degree of agglomeration) led to both different onset temperatures for FS and final pellet microstructures. However, such differences in terms of dense pellet properties comparing commercial precursors and as-synthesized precursors are much less evident in the case of FS adoption rather than conventional sintering.

As an example, by considering GDC10 samples sintered by FS starting from co-precipitated powders, a flash event is always evident, occurring within one minute. The onset temperature is always lower than 700 °C, and it decreases by about 150 °C upon a corresponding increase of the field from 75 to 150 V/cm.

It is important to note that flash sintering does not modify the crystalline phase evolution of ceria-based materials, as all sintered pellets exhibit only cubic fluorite-like-doped ceria [4, 6].

The flash sintered samples relative density is generally very high, within the range of 98–100%. Interestingly, the sample flash sintered under 75 V/cm exhibits the highest relative density, i.e. around 99.5%. For GDC10, the relative density slowly decreases with the increase of the applied electric field; this effect can account for the

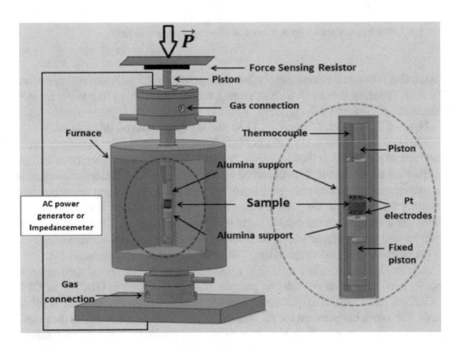

Fig. 5.1 Schematic representation of FS experimental setup

Fig. 5.2 SEM micrographs of sample sintered by flash sintering cycle at lower (**a**) and higher (**b**) magnification, respectively

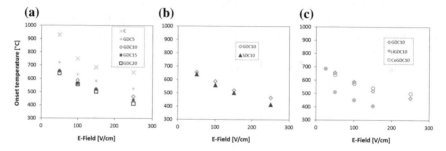

Fig. 5.3 Onset temperature for FS as a function of the applied electric field

correlation between sintering temperature and applied field; i.e. the onset temperature for flash sintering decreases with increasing fields, thus leading to a lower furnace temperature required by the process.

Figure 5.2 shows two exemplary SEM micrographs taken on this sample.

The very small grain size shown in Fig. 5.2 has a beneficial role not only upon the mechanical properties of the electrolyte, but also upon its electrical behavior. In fact, in the case of nanocrystalline-doped ceria, the presence of submicroscopic grains characterized by a much larger grain boundary area results in diluting the solutes and impurity segregation at the grain boundaries. In addition, the very limited treatment times and temperatures can help to avoid segregation formation. This should lead to an enhanced grain boundary ionic conduction. Therefore, all microstructure features of the flash sintered pellets fulfill the IT-SOFCs electrolytes requirements [5].

All the obtained results for GDC10 extend to other similar ceria-based systems too. In particular, even considering samples C, GDC5, GDC15, GDC20, SDC10, LiGDC10, and CoGDC10, the flash sintering phenomenon was observed in all of them. The onset temperature for FS, defined as the temperature at which the power supply switches from voltage to current control, is shown in Fig. 5.3 as a function of the electric field. As already mentioned, FS temperature decreases when higher electrical field is applied [2]. Anyway, the flash sintering temperature is always well below the values typically required for conventional sintering (i.e. about 1500 °C).

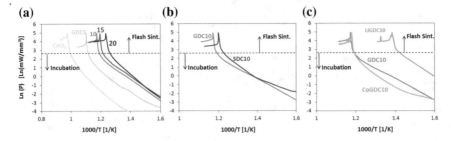

Fig. 5.4 Specific power dissipation as a function of furnace inverse temperature ($E = 100$ V/cm)

The onset temperature is influenced by the presence of dopants and sintering aids. Specifically, larger Gd and Sm doping accounts for lower FS temperature; the reduction is particularly significant (i.e. about 200 °C) up to 10 mol% doping. As for the sintering aids, Li$_2$O drastically reduces FS temperature by more than 100 °C, whereas CoO does not seem to significantly affect the sintering behavior (the difference is always lower than 35 °C).

It is well known that the onset temperature for flash sintering is related to the specific power dissipation (P) trend. Figure 5.4 shows how P changes with the furnace temperature for samples subjected to 100 V/cm.

In all cases, a deviation from linearity occurs for P~15 mW/mm^3, in agreement with previous results [5, 8]. The flash phenomenon is observed ~30–60 s (10–20 °C) after such power dissipation value is reached.

This behavior is very likely due to the fact that, once the threshold specific power is reached, the sintering sample is no more able to dissipate the internal heat generated by Joule effect and it undergoes to a rapid and uncontrolled heating [2]. After that, the system switches from voltage to current control and the flash event occurs. It can be also pointed out that, during the incubation (i.e. when the system is under voltage control), the specific power dissipation increases with Gd or Sm content up to 10 mol%, while further dopant additions do not lead to significant changes. Furthermore, Li addition drastically increases power dissipation during FS incubation, while no significant effects can be observed in the Co-containing material.

By comparing Fig. 5.3 and Fig. 5.4, the onset temperature is lower for specimens characterized by higher power dissipation during FS incubation. When the system is under voltage control (FS incubation stage), power dissipation linearly increases with green body conductivity; in other words, green pellets with higher conductivity are characterized by larger P and lower onset temperature (under the same field condition). Therefore, by controlling the type of dopant and its content, it is possible to alter green body conductivity, further decreasing the onset sintering temperature.

An analytical relation between the onset temperature for FS (T_o) and the electrical parameters of the system was developed by Dong et al. [9]:

$$\ln\left(\frac{E^2}{T_o^4}\right) = \frac{E_a}{RT_{on}} + B \tag{5.1}$$

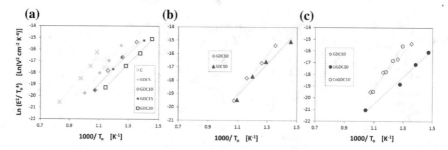

Fig. 5.5 Correlation between electric field (E) and onset temperature (T_o) in the $\ln(E^2/T_o^4)$ versus $1000/T_o$ plots, the dashed lines representing Eq. (5.1) interpolation

where E_a is the activation energy for conduction in the green body, R is the universal gas constant, and B is a constant depending on geometrical and irradiative parameters of the system.

The results recorded during FS experiments are reported in Fig. 5.5. The data follow a linear behavior ($R^2 \leq 0.976$), according to Eq. (5.1). The activation energy for conduction in the green specimen can be estimated from the slope of the plots in Fig. 5.5, and it can be compared with the activation energy measured from the power dissipation plots during the incubation of flash sintering (see Fig. 5.4). E_a values range between 0.93 and 1.65 eV, being in quite good agreement with the activation energies measured in dense electrolytes [10]. Finally, the results obtained with the two methods are similar and this result supports the validity of the model developed for estimating the onset temperature for flash sintering, providing a further validation of Eq. (5.1).

In the presence of Li, the large FS temperature decrease plays a negative role in the material densification. Probably, the onset temperature is too low and the material cannot attain a satisfactory densification. Its microstructure observed on the fracture surface of the sample flash sintered at 100 V/cm is shown in (Fig. 5.6 a, b).

The grain size is homogenous (around 500 nm) and a well-distributed nanosized porosity is present, thus determining a relatively low density of around 67%. There-fore, according to these results, the use of Li as sintering aid in flash sintering of Gd-doped ceria ceramics appears not only useless, but also even counterproductive. As shown in Fig. 5.6, the grain size in the other FS samples is definitely smaller as well as the pores size. It seems that coarsening phenomena rather than densification phenomena occurred during flash sintering in LiGDC10, the effect being stronger at higher applied fields.

CoGDC10, GDC10, and SDC10 exhibit similarities in terms of grain size and grain texture of the well-densified areas. However, CoGDC10 and SDC10 are not completely homogeneous and they are formed by nearly fully dense regions alter-nated with porous regions (see Fig. 5.6c, g). Conversely, GDC is very well densified (98–99% of relative density), with very fine and homogeneous grain size (i.e. around 200–300 nm), as it is clear from its microstructure reported in (Fig. 5.6e, f).

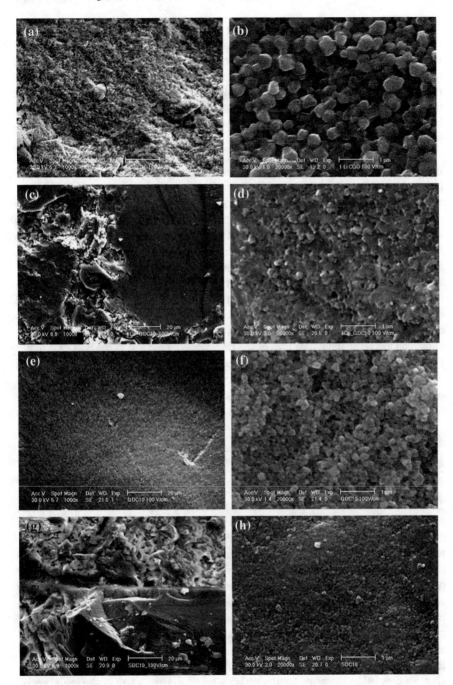

Fig. 5.6 SEM micrographs of flash sintered LiGDC10 (**a**, **b**), CoGDC10 (**c**, **d**), GDC10 (**e**, **f**), and SDC10 (**g**, **h**)

FS has been successfully proven to be an effective method to fully sinter GDC specimens starting from commercial-grade powders [7], even if a presintering step is required to achieve satisfactory green densities. By considering a series of GDC20 samples, the flash event always occurred at temperatures decreasing with the applied field (ranging from 15 to 70 V/cm). An evident reduction in grain growth with the increase of the electric field is confirmed in this case too; in particular, at sufficiently high values of the applied electric field, the average grain size of sintered pellets tends toward values being very close to the related raw material size (i.e. 0.3–0.5 μm).

Figure 5.7 shows SEM micrographs of such flash sintered GDC20 pellets, highlighting the average grain size reduction from 0 V/cm (Fig. 5.7a, conventional sintering) to 70 V/cm (Fig. 5.7f).

According to these findings, nanocrystalline Gd- and Sm-doped ceria powders synthesized by co-precipitation method can be successfully consolidated by flash sintering at temperature between 600 and 900 °C under electric field ranging from 5 to 250 V/cm. Quite surprisingly, the addition of Li_2O and CoO as sintering aids does not improve the achieved densification.

Definitely, flash sintering process can be applied to ceria-based systems to produce full dense pellets in a few seconds and at reduced temperature, as it has also been proved for YSZ [1, 3]. Moreover, cheap raw materials (i.e. co-precipitated powders or even commercial powders) are suitable for FS without any use of sintering aids, thus lowering the whole cost of the final material as well as the cost of primary energy for the sintering process. Finally, the reduced sintering temperatures required could lead to an innovative way to simultaneously sinter the anode–electrolyte nano-/microstructure, i.e. avoiding the possible collapse of the anode nanostructure eventually caused by the high sintering temperatures required by conventional processes.

5.2   Fast Firing (FF)

To define even more performant sintering cycles, the effect of extremely high heating rates has been analyzed, bearing in mind that densification and grain growth are competitive phenomena during sintering. In fact, upon sintering of polycrystalline ceramics, both thermodynamics and kinetics contribute to the competition of grain/pore growth (coarsening) and densification processes. The specific surface energy of grain boundaries as well as thermally activated grain boundary mobility and geometric parameters controls the grain growth kinetics. Densification requires diffusion of material into the intergranular pores (necking), while the pressure of residual gases in the pores opposes both grain boundary migration and neck growth, as determined by the gas/solid surface energy [11]. During the initial stages of sintering, both grain growth and densification are inversely proportional to the grain size. Studies in the literature have noted that when the grain size of the initial ceria-based system is in the nanometric range, densification is improved and the sintering temperature can be lowered. From this point of view, the heating rate is a crucial parameter to control the coarsening/densification ratio when the two processes have

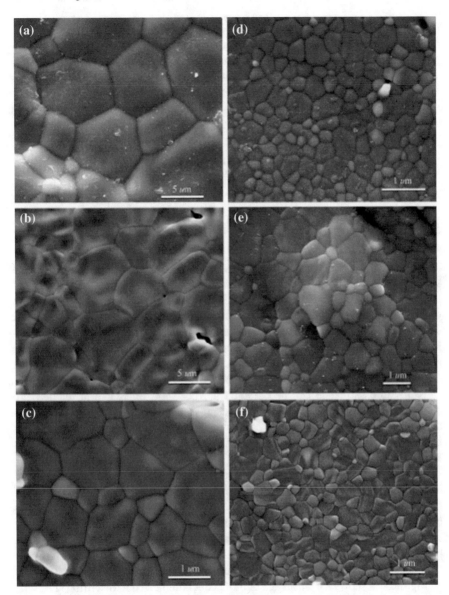

Fig. 5.7 SEM micrographs of flash sintered GDC20 samples at different applied electric fields: 0 V/cm (**a**), 20 V/cm (**b**), 30 V/cm (**c**), 40 V/cm (**d**), 50 V/cm (**e**), 70 V/cm (**f**)

different activation energies. In particular, if the activation energy for grain growth is lower than that for densification, high heating rate has a beneficial effect for activating the densification mechanisms without significant prior coarsening phenomena. This leads to extremely high sintering rates induced by the small grain size. This kind of process is referred as fast firing [12].

More practically, a "fast firing" (FF) is also defined as a sintering duration from ambient to ambient within 180 min or less.

The main advantages of a FF sintering protocol consist in significantly reducing energy consumption per payload, shorter production lead times, potential stabilization at room temperature of metastable phases, reduced grain size.

5.2.1 Rapid Densification of Gd-/Sm-Doped Ceria

The fast firing (FF) sintering process can be applied on doped ceria systems as a rapid sintering protocol starting from nanometric and highly reactive powders carried out in a preheated tubular furnace at an optimal peak temperature (depending on different doping loads) and reduced dwell times (i.e. in the order of several minutes or less). All samples subjected to FF have to be preheated to outgas them and then quickly moved into the central part of the furnace for different soaking times. Furnace temperatures can range from 900 to 1100 °C for SDC [13] and up to 1500 °C for GDC or undoped Ceria [11], while soaking times vary from 15 to 300 s.

Noticeable shrinkage can be obtained in very short times and at temperature significantly lower than the typically needed to sinter ceria-based systems. This sintering behavior is related to the combined effect of extremely high heating rates and small grain size; i.e. nanometric powders are required to successfully apply this kind of sintering protocol.

Moreover, the extremely high sintering rates obtained upon fast firing are somehow compatible with those observed in other sintering processes like flash sintering (FS). During FS, in fact, sample temperature drastically increases in few seconds because of the Joule heating generated by the current flowing through the green specimen. High heating and sintering rates are common points between FF and FS. Nevertheless, being FF "field-less," it is simpler than FS with less geometrical constraints and with no problems related to current concentration phenomena.

Figure 5.8a shows the shrinkage of SDC pellets subjected to fast firing (FF) with variable duration and at different maximum temperatures.

In this specific case, as shown in Fig. 5.8a, the densification mechanisms of SDC nanopowders are already activated at 900–1000 °C. This means that such electrolyte can be coupled with anode materials different from those conventionally used (i.e. NiO), like copper-based anodes that are more advantageous in terms of degradation properties and toxicity [14].

Grain size evolution as a function of time and temperature is shown in Fig. 5.8b. In almost all cases, the grain size (evaluated by XRD on sintered pellets after milling) is far below 100 nm, even when shrinkage is as high as 19%.

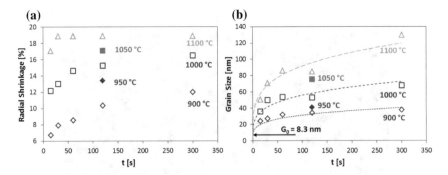

Fig. 5.8 Radial shrinkage upon fast firing (**a**) and grain size in the sintered bodies (**b**) as a function of sintering time and temperature

The grain size evolution was investigated by using the well-known relationship for grain coarsening [12]:

$$G^m = G_0^m + k_0 \exp(-Q/RT) \qquad (5.2)$$

where G is the grain size, G_0 the initial grain dimension (equal to 8.3 nm, as from XRD measurement on the as-synthesized powder), m a dimensionless exponent, T the absolute temperature, R the universal gas constant, Q the activation energy for grain boundary motion, k_0 a pre-exponential constant, and t the treating time. The experimental data were fitted using the least square method, and the best fits, shown in Fig. 5.8b, were obtained using m $= 4.1$ and $Q = 298$ kJ/mol. The estimated exponent (m) is very close to 4, and it is in good agreement with the literature data for grain growth in ceramic systems [12]. The activation energy, i.e. 298 kJ/mol, could be related to the diffusion of cations (Ce^{4+} or Sm^{3+}) through the grain boundary, being the activation energy for oxygen diffusion much lower in the present material (~53 kJ/mol [10]).

The reported data allow defining that shrinkage *vs* grain size trajectory is sensitive to the treating temperature. Therefore, it is possible to obtain a denser material with the same grain size, by increasing the sintering temperature within the range 900–1100 °C. In other words, the same sintering level is achieved with smaller grains for samples treated at higher temperatures.

This behavior is an effect of the different activation energies for coarsening and densification, the second being higher. Hence, at low temperature, coarsening mechanisms are more likely to occur with respect to the densification ones. For this reason, a high heating rate treatment is very effective to improve the densification behavior of nanograined SDC. The fast heating process allows going very quickly through the temperature region in which grain growth is already active, whereas sintering is not operative yet.

It is also possible to state that the optimum combination of treating temperature and time is represented by the sample sintered at 1100 °C for 30 s. Such processing

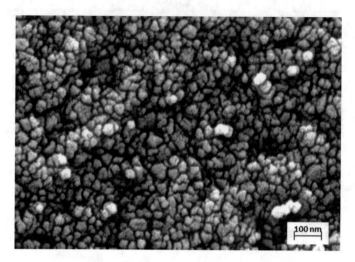

Fig. 5.9 FE-SEM micrograph of the free surface of sample fast fired at 1100 °C for 30 s

parameters allow obtaining a dense material with very limited grain size (71 nm). Longer treatments do not enhance the densification (already completed in 30 s), but, conversely, they induce grain coarsening phenomena.

The surface of fast fired samples was observed by FE-SEM to visualize the densification and grain growth phenomena. Figure 5.9 shows an exemplary micrograph of SDC sample treated at 1100 °C for 30 s.

The material is well densified, and almost no pores are detectable. Moreover, the grain size is still nanometric (far below 100 nm) and this very limited grain size can have beneficial effect both on electrical and mechanical properties of the ceramic sintered body [15].

However, bulk density of the considered SDC samples was estimated to reach about 94% at the highest and, very likely, higher temperatures are needed to obtain denser materials.

In [11], it is reported that FF of both GDC and undoped ceria requires peak temperatures as high as 1350 °C to achieve relative density values > 96%. In particular, Gd doping content has a significant role in the determination of the optimal sintering temperature (at soaking times <1 min), as this peak temperature increases with Gd load via a parabolic dependence and at its maximum relative densities >98% can be obtained (see Fig. 5.10).

The parabolic dependence between the achieved density and FF temperature could be explained by the presence of the two competing and thermally activated processes occurring upon sintering: densification and pore/grain growth. A similar trend has been reported for SDC too [13], thus confirming that at higher temperatures pore growth becomes dominant and the final pellet is less dense. Furthermore, it is inter-

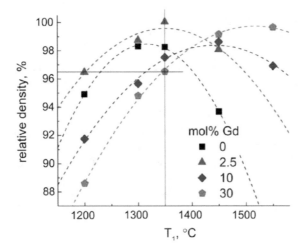

Fig. 5.10 Relative densities of undoped and Gd-doped ceria pellets as a function of peak temperature (T_1) in FF sintering protocol

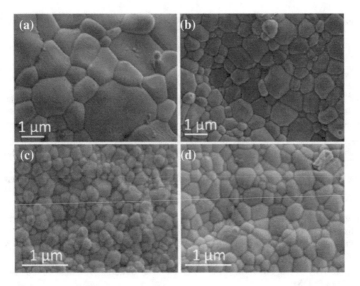

Fig. 5.11 SEM micrographs of samples fast fired at 1350 °C for about 1 min, undoped ceria (**a**), 2.5 mol% Gd (**b**), 10 mol% Gd (**c**), and 30 mol% Gd (**d**)

esting to note from Fig. 5.10 that 1350 °C represents a good compromise as FF temperature for doped ceria system over a wide range of compositions, being the final pellet relative density always higher than 96%.

Figure 5.11 shows some exemplary SEM micrographs of ceria-based pellets fast fired at 1350 °C for about 1 min at Gd doping contents ranging from 0 to 30%.

All samples shown in Fig. 5.11 exhibit a very dense and homogeneous microstructure, with virtually no residual porosity. Conversely, the average grain size is inversely dependent on doping load, as previously reported for Gd-doped ceria too [16]. The slightly higher value of 30GDC (Fig. 5.11d), average grain size is very likely related to some residual agglomeration of the corresponding parent powder, as reported in [11].

According to such encouraging results, FF represents a very promising sintering route for nanograined-doped ceria powders, allowing to drastically reduce sintering times and temperatures and to produce dense pellets with an excellent microstructure. However, FF can be applied to samples with limited dimensions to avoid significant thermal stresses possibly causing cracks and failures in bigger samples. That said, many ceramic components have suitable geometry for being consolidated by FF such as IT-SOFCs electrolytes.

References

1. R. Raj, M. Cologna, A.L.G. Prette, V.M. Sglavo, Methods of flash sintering. United States of America Brevetto US20130085055 A1, 4 Apr 2013
2. E. Zapata-Solvas, D. Garcia, A. Rodriguez, R.I. Todd, Ultra-fast and energy-efficient sintering of ceramics by electric current concentration. Sci. Rep. **5**, (2015)
3. M. Yu, S. Grasso, R. Mckinnon, T. Saunders, M.J. Reece, Review of flash sintering: materials, mechanisms and modelling. Adv. Appl. Ceram. **116**(1) 24–60 (2016)
4. T. Jiang, Z. Wang, J. Zhang, X. Hao, D. Rooney, Y. Liu, W. Sun, J. Qiao, K. Sun, Understanding the flash sintering of rare-earth-doped ceria for solid oxide fuel cell. J. Am. Ceram. Soc. **98**(6), 1–7 (2015)
5. M. Biesuz, G. Dell'Agli, L. Spiridigliozzi, C. Ferone, V.M. Sglavo, Conventional and field-assisted sintering of nanosized Gd-doped ceria synthesized by co-precipitation. Ceram. Int. **42**(10), 11766–11771 (2016)
6. L. Spiridigliozzi, M. Biesuz, G. Dell'Agli, E.D. Bartolomeo, F. Zurlo, V.M. Sglavo, Microstructural and electrical investigation of flash-sintered Gd/Sm-doped ceria. J. Mater. Sci. **52**(12), 7479–7488 (2017)
7. X. Hao, Y. Liu, Z. Wang, J. Qiao, K. Sun, A novel sintering method to obtain fully dense gadolinia doped ceria by applying a direct current. J. Power Sources **210**, 86–91 (2012)
8. J.A. Downs, V.M. Sglavo, electric field assisted sintering of cubic zirconia at 390 °C. J. Am. Ceram. Soc. **96**(5), 1342–1344 (2013)
9. Y. Dong, I.W. Chen, Predicting the onset of flash sintering. J. Am. Ceram. Soc. **98**(8), 2333–2335 (2015)
10. M. Mogensen, N.M. Sammes, Physical, chemical and electrochemical properties of pure and doped ceria. Solid State Ionics **129**, 63–94 (2000)
11. N. Yavo, A. Nissenbaum, E. Watchel, T. Shaul, O. Mendelson, G. Kimmel, S. Kim, I. Lubomirsky, O. Yeheskel, Rapid sintering protocol produces dense ceria-based ceramics. J. Am. Ceram. Soc., p. Article in press (2018)
12. M. N. Rahaman, *Ceramic Processing and Sintering*. (CRC Press, 2003)
13. M. Biesuz, L. Spiridigliozzi, M. Frasnelli, G. Dell'Agli, V.M. Sglavo. Mater. Lett. **190**, 17–19 (2017)
14. A. Azzolini, V.M. Sglavo, J.A. Downs, Production and Performance of Copper-based Anode-supported SOFCs, in *ECS Conference on Electrochemical Energy Conversion & Storage with SOFC*, Glasgow, 2015

15. D. Xu, H.K. Li, Y.J. Zhou, Y. Gao, D.T. Yang, S.F. Xu, The effect of NiO addition on the grain boundary behavior and electrochemical performance of Gd-doped ceria solid electrolyte under different synthesis conditions. J. Eur. Ceram. Soc. **37**(1), 419–425 (2017)
16. N. P. Bansal, P. Singh, *Advances in Solid Oxide Fuel Cells V*. (Wiley, 2010)

Appendix A
Theoretical Background

A.1 Ionic Conduction in Solids

Parameters influencing conductivity in the solid-state include the concentration of charge carriers, crystal temperature, availability of vacant-accessible sites, defects density into the crystal, and ease with which an ion to jump toward different sites.

Hence, ionic conductivity can be expressed as electronic conductivity, according to:

$$\sigma = n \cdot q \cdot \mu \tag{A.1}$$

Where n is the number of charge carriers per unit volume, q is their unit charge, and μ is their mobility, being a measure of the drift velocity under a constant electric field. Equation (1) is the general equation defining conductivity for all conducting materials. A closer look to the hopping model is useful to better understand ionic conduction and how a solid is a better ionic conductor than another one.

Firstly, an electric current in an ionic conductor is carried by the defects. By considering crystals, in which the ionic conductivity is based on either vacancy or interstitial mechanisms, the charge carriers' concentration (n) is closely related to defects concentration into the crystal, whereas μ is referred to these defects mobility.

The required energy to let an ion jump to another site is represented by the activation energy E_a, being it a phenomenological value. Activation energy is typically used to indicate the free energy barrier that an ion must overcome to successfully jump among different sites.

In solids, defect moves through crystals in a diffusive way, as the ionic charge carriers mobility follow the Nernst–Einstein equation:

$$\mu = \frac{qD}{kT} \tag{A.2}$$

© The Author(s), under exclusive license to Springer Nature Switzerland AG 2018
L. Spiridigliozzi, *Doped-Ceria Electrolytes*, SpringerBriefs in Applied Sciences
and Technology, https://doi.org/10.1007/978-3-319-99395-9

Where q is the carrier charge, T is the absolute temperature, k is the Boltzmann constant, and D is the diffusion constant. D can be expressed as:

$$D = D_0 \exp(-Q)/kT \tag{A.3}$$

Where D_0 is a material-dependent constant and Q is the activation energy of the transport process (measured in eV).

For an ion to move toward an energetically equivalent unoccupied neighboring site, the phenomenological expression of the ionic conductivity therefore can be expressed as:

$$\sigma = n \cdot q \cdot \mu = \sigma_0 \cdot e^{\frac{E_a}{kT}} \tag{A.4}$$

According to Equation (3), the pre-exponential term σ_0 incorporates n and q, as well as the information about attempt frequency and jump distance. The activation energy E_a incorporates the motion enthalpy and the energy associated with n mobile carriers' creation (including trapping energy of defects). The latter contribution to E_a could be temperature dependent, consequently contributing in a significant manner to the pre-exponential factor. According to the theory of ionic charge carriers random walk, the pre-exponential factor can be differently expressed by giving to the conductivity equation a new form as follows:

$$\sigma = N \frac{q^2}{kT} \gamma c (1 - c) a_0^2 \cdot v_0 \exp\left\{\frac{-\Delta S_m}{k}\right\} \exp\left\{\frac{-E_a}{kT}\right\} \tag{A.5}$$

Where a_0 is the jump distance to nearest neighbor equivalent sites, v_0 is the jump attempt frequency, $c = n/N$ is the fraction of mobile ion species on N equivalent sites per cm^3, ΔS_m is the entropy associated with ion migrations, and γ is the correlation factor which depending on conductivity mechanism.

Parameters influencing conductivity the most are c, ΔS_m, and E_a although a certain interdependence between them exists.

A.1.1 Ion Conduction Mechanisms in Solid Electrolytes

As stated in Sect. 6.1, crystal point defects are responsible for eventual atoms/ions movements through a solid-state structure. For an ideally perfect structure, in fact, it would be difficult to predict how such movements (either lattice diffusion or ionic conductivity) could take place.

Schottky and Frenkel defects (see Figure A.1) are the two kinds of defects responsible for ion transport in crystals, belonging to the class of point defects.

Schottky defects consist in a pair of ions (both cation and anion) disappearing from the structure and leaving their position vacant. Frenkel defects consist in

Figure A.1 Schematic representation of Schottky and Frenkel defects

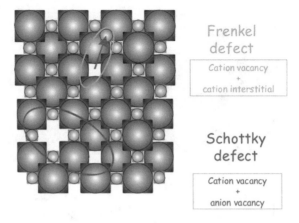

single ions wandering from their regular position to interstitial sites. Hence, an ion species mobility into a stable crystal requires point defects that typically move according to one of the three elementary jump mechanisms.

Ions moving to an interstitial site (creating a Frenkel defect) can subsequently jump to another neighboring interstitial site and so on, resulting in long-term motion and travelled distances. Such jumping mechanism is labeled as interstitial migration. Two types of interstitial migration exist: the direct interstitial mechanism and the indirect interstitial mechanism. The former consists in defects directly jumping into another interstitial site, whereas the latter is based on defects pushing an ion toward an adjacent empty interstitial site. Because the interstitial sites in most crystalline solids are small, interstitial migration is a high-energy process relatively uncommon.

In a vacancy-based mechanism, both Frenkel and Schottky defects result in vacant crystal sites, and consequently, any ion in the immediate neighborhood can jump to one of those vacant sites. Such process leaves the site previously occupied by the ion vacant and available to host another ion, and so on. Practically, the whole process leads to continuous ions movement through the solid, thus giving rise to its ionic conductivity, being vacancies present in a significant concentration in all crystalline materials.

A.2 Sintering

Solid-state sintering is a rather complex process, usually defined as the final step of ceramic production, consisting in converting powders into dense solids upon heating. The sintering process is therefore crucial during a ceramic production cycle, since the obtained microstructure defining the material desired properties strongly depends on it.

Solid-state sintering can be roughly considered occurring in three subsequent phases, as shown in Figure A.2.

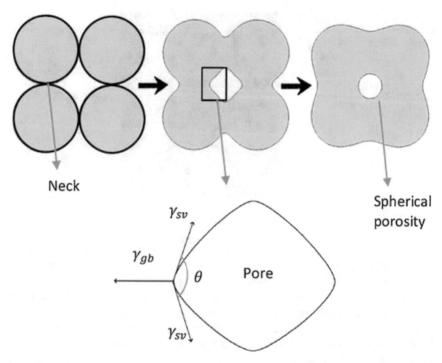

Figure A.2 Schematic representation of solid-state sintering process, along with pore formation

During the first sintering stages, necks formation and growth are the processes taking place in compacted powders. Pores are interconnected, and pore shapes are heterogeneous and irregular. As the intermediate stage goes by, neck growth becomes significant and pore channels tend to be cylindrical in shape. Pores and particles form an intersecting network, the achieved density consequently increasing. Curvature gradients across the necks are responsible for mass flow, as the driving force upon this stage is the interfacial energy. Small neck size accounts for higher curvature gradients, thus leading to faster sintering. As the neck grows, the curvature gradient consequently decreases and the sintering rate does too. Pore rounding can also occur in this intermediate phase, without any shrinkage or variation in pore total volume, being only a change in pore shape (possibly becoming isolated from the pore channel). Upon prolonged sintering, pore channels become unstable, gradually pinching off and closing, and a network of isolated pores mixed to a skeleton of solid particles is formed. In this final stage, grain boundaries migration by grain growth phenomena occurs, along with mainly the whole material shrinkage. Moreover, pores are no longer interconnected, the residual porosity being located either at the grain boundaries or within the grains. Upon this stage, densification process proceeds at very slow rates and residual porosity exists even after long sintering times.

As already mentioned, the driving force for the initial stage of solid-state sintering is the excess of surface free energy. Upon sintering, the powder compact aims to minimize its surface energy by diffusing material from different areas through different transport mechanisms. Upon intermediate and final stages of sintering, grain growth takes place, leading to significant reduction of grains total number.

Sintering of crystalline materials is accounted for several mechanisms, including vapor transport, surface diffusion, lattice diffusion, grain boundary diffusion, lattice diffusion, and plastic flow. However, one mechanism usually has control over the others upon different sintering stages. Figure A.3 shows a schematic representation of matter transport paths between sintering particles.

Commonly, a distinction between densifying and nondensifying mechanisms is made. Nondensifying mechanisms (1, 2, and 3 in Figure A.3) produce microstructural changes without causing shrinkage, whereas densifying mechanisms (4, 5, and 6 in Figure A.3) remove material from grain boundary regions consequently causing shrinkage.

Vapor transport, surface diffusion, and lattice diffusion from particle surfaces to necks lead to neck growth and particles coarsening. Grain boundary diffusion and lattice diffusion from grain boundaries to necks are the most important mechanisms in polycrystalline ceramics, allowing neck growth as well as shrinkage. Finally, plastic flow is mainly governed by dislocation motion causing neck growth and densification through particles deformation (i.e. creep), and it is the most important densifying mechanism in metal powders.

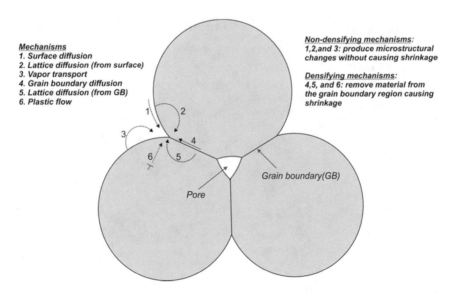

Figure A.3 Material transport mechanisms operating upon solid-state sintering

Bibliography

1. N.A. Razik, Precise lattice constants determination of cubic crystals from x-ray powder diffractometric measurements. Appl. Phys. A **37**(3), 187–189 (1985)
2. P. Murray-Rust, Crystallography open database. SOLSA Project, [Online]. Available: www.crystallography.net.
3. J.E. Bauerle, Study of solid electrolyte polarization by a complex admittance method. J. Phys. Chem. Solids **30**(12), 2657–2670 (1969)
4. E.C.C. Souza, Electrochemical properties of doped ceria electrolyte under reducing atmosphere: bulk and grain boundary. J. Electroceram. **31** (1–2) (2013)
5. E.C.C. Souza, W.C. Chueh, W. Jung, E.N.S. Muccillo, S.M. Haile, Ionic and electronic conductivity of nanostructured, samaria-doped ceria. J. Electrochem. Soc. **129** (5) 127–135 (2012)
6. G. Dell'Agli, G. Mascolo, M.C. Mascolo, C. Pagliuca. Crystallization-stabilization mechanism of yttria-doped zirconia by hydrothermal treatment of mechanical mixtures of zirconia xerogel and crystalline yttria. J. Cryst. Growth, 255–265 (2005)
7. G. Dell'Agli, G. Mascolo, M.C. Mascolo, C. Pagliuca, Drying effect on thermal behaviour and structural modification of hydrous zirconia gel. J. Am. Ceram. Soc. 3375–3379 (2008)
8. C. Suryanarayana, M.G. Norton, X-ray diffraction, a practical approach (Springer, 1998)
9. N. Sonoyama, S.G. Martin, U. Amador, N. Imanishi, M. Ikeda, N. Erfua, H. Tanimura, A. Hiranod, Y.Takeda, O. Yamamoto, Crystal structure and electrical properties of magnesia co-doped scandia stabilized zirconia. J. Electrochem. Soc. 1397–1401 (2015)
10. N. Mahato, A. Banerjee, A. Gupta, S. Omar, K. Balani, Progress in material selection for solid oxide fuel cell technology: a review. Prog. Mater. Sci. 141–337 (2015)
11. T.I. Politova, J.T.S. Irvine, Investigation of scandia–yttria–zirconia system as an electrolyte material for intermediate temperature fuel cells—influence of yttria content in system $(Y_2O_3)_x(Sc_2O_3)_{(11-x)}(ZrO_2)_{89}$. Solid State Ionics **168** (1–2), 153–165 (2004)
12. J. Jacob, R. Bauri, One step synthesis and conductivity of alkaline and rare earth co-doped nanocrystalline CeO_2 electrolytes. Ceram. Int. **41**, (5), 6299–6305 (2015)
13. H. Li, C. Xia, M. Zhu, Z. Zhou, G. Meng, Reactive Ce0.8Sm0.2O1.9 powder synthesized by carbonate coprecipitation: sintering and electrical characteristics. Acta Mater. **54**(3), 721–727 (2006)
14. R. Peng, C. Xia, Q. Fu, G. Meng, D. Peng, Novel intermediate temperature ceramic fuel cells with doped ceria-based composite electrolytes. Solid State Ionics **148**(3–4), 533–537 (2002)
15. S. Lubke, H.D. Wiemhofer, Electronic conductivity of Gd-doped ceria with additional Pr-doping. Solid State Ionics **117** (3–4), 229–243 (1999)

© The Author(s), under exclusive license to Springer Nature Switzerland AG 2018
L. Spiridigliozzi, *Doped-Ceria Electrolytes*, SpringerBriefs in Applied Sciences and Technology, https://doi.org/10.1007/978-3-319-99395-9

16. M.L.D. Santos, R.C. Lima, C.S. Riccardi, Preparation and characterization of ceria nanospheres by microwave-hydrothermal method. Mater. Lett. **62**(30), 4509–4511 (2008)

17. L. Lutterotti, M. Bortolotti, G. Ischia, I. Leonardelli, H.R. Wenk, Rietvield texture analysis from diffraction images. Cryst. Mater. **26**, 125–130 (2007)

18. S.G. Ullattil, P. Periyat, Sol-Gel synthesis of Titanium Dioxide in Sol-Gel materials for energy, environment and electronic applications, ed. by S.C. Pillai, S. Hehir (2017), pp. 271–283

19. K. Zupan, M. Marinsek, S. Pejovnik, J. Macek, K. Zore, Combustion synthesis and the influence of precursor packing on the sintering properties of LCC nanopowders. J. Eur. Ceram. Soc. **24** (6), 1935–1939 (2004)

20. H.W. Wang, D.A. Hall, F.R. Sale, A thermoanalytical study of the metal nitrate—EDTA precursors for lead zirconate titanate ceramic powders. J. Thermal Anal. **41**, 2–3 (1994)

21. P. Courty, H. Ajot, C. Marcilly, B. Delmon, Oxydes mixtes ou en solution solide sous forme très divisée obtenus par décomposition thermique de précurseurs amorphes. Powder Technol. **7**(1), 21–38 (1973)

22. X. Li, Z. Feng, J. Lu, F. Wang, M. Xue, G. Shao, Synthesis and electrical properties of $Ce_{1-x}Gd_xO_{2-x/2}$ (x = 0.05–0.3) solid solutions prepared by a citrate–nitrate combustion method. Ceram. Int. **38**(4), 3203–3207 (2012)

23. A. Sin, Y. Dubitsky, A. Zaopo, Preparation and sintering of $Ce_{1-x}Gd_xO_{2x/2}$ nanopowders and their electrochemical and EPR characterization. Solid State Ionics **175**(1–4), 361–366 (2004)

24. G. Accardo, C. Ferone, R. Cioffi, D. Frattini, L. Spiridigliozzi, G. Dell'Agli, Electrical and microstructural characterization of ceramic gadolinium-doped ceria electrolytes for ITSOFCs by sol-gel route. J. Appl. Biomater. Funct. Mater. **14** (1), 35–41 (2016)

25. W.J. Bowman, J. Zhu, R. Sharm, P. Crozier, Electrical conductivity and grain boundary composition of Gd-doped and Gd/Pr co-doped ceria. Solid State Ionics **272**, 9–17 (2015)

26. G. Dell'Agli, G. Mascolo, M.C. Mascolo, C. Pagliuca, Weakly-agglomerated nanocrystalline $(ZrO_2)_{0.9}(Yb_2O_3)_{0.1}$ powders hydrothermally synthesized at low temperature. Solid State Sci. **8**(9), 1046–1050 (2006)

27. G. Dell'Agli, G. Mascolo, Agglomeration of 3 mol% Y±TZP powders sythesized by hydrothermal treatment. J. Eur. Ceram. Soc. **21**, 29–35 (2001)

28. Z. Lu, J. Hardy, J. Templeton, J. Stevenson, D. Fisher, N. Wu, A. Ignatiev, Performance of anode-supported solid oxide fuel cell with thin bi-layer electrolyte by pulsed laser deposition. J. Power Sources **210**, 292–296 (2012)

29. S. Chengz, C. Chatzichristodoulou, M. Sogaard, A. Kaiser, P.V. Hendriksen, Ionic/electronic conductivity, thermal/chemical expansion and oxygen permeation in Pr and Gd Co-Doped Ceria PrxGd0.1Ce0.9-xO1.95-δ. J. Electrochem. Soc. **164** (13), 1354–1367